Sustainable Civil Infrastructures

Sustainable Infrastructure impacts our well-being and day-to-day lives. The infrastructures we are building today will shape our lives tomorrow. The complex and diverse nature of the impacts due to weather extremes on transportation and civil infrastructures can be seen in our roadways, bridges, and buildings. Extreme summer temperatures, droughts, flash floods, and rising numbers of freeze-thaw cycles pose challenges for civil infrastructure and can endanger public safety. We constantly hear how civil infrastructures need constant attention, preservation, and upgrading. Such improvements and developments would obviously benefit from our desired book series that provide sustainable engineering materials and designs. The economic impact is huge and much research has been conducted worldwide. The future holds many opportunities, not only for researchers in a given country, but also for the worldwide field engineers who apply and implement these technologies. We believe that no approach can succeed if it does not unite the efforts of various engineering disciplines from all over the world under one umbrella to offer a beacon of modern solutions to the global infrastructure. Experts from the various engineering disciplines around the globe will participate in this series, including: Geotechnical, Geological, Geoscience, Petroleum, Structural, Transportation, Bridge, Infrastructure, Energy, Architectural, Chemical and Materials, and other related Engineering disciplines.

More information about this series at http://www.springer.com/series/15140

Sherif El-Badawy · Ragaa Abd El-Hakim
Editors

Recent Developments in Pavement Design, Modeling and Performance

Proceedings of the 2nd GeoMEast International Congress and Exhibition on Sustainable Civil Infrastructures, Egypt 2018 – The Official International Congress of the Soil-Structure Interaction Group in Egypt (SSIGE)

Springer

Editors
Sherif El-Badawy
Public Works Engineering Department,
Faculty of Engineering
Mansoura University
Mansoura, Egypt

Ragaa Abd El-Hakim
Tanta University
Tanta, Egypt

ISSN 2366-3405 ISSN 2366-3413 (electronic)
Sustainable Civil Infrastructures
ISBN 978-3-030-01907-5 ISBN 978-3-030-01908-2 (eBook)
https://doi.org/10.1007/978-3-030-01908-2

Library of Congress Control Number: 2018957406

This Springer imprint is published by the registered company Springer Nature Switzerland AG
The registered company address is: Gewerbestrasse 11, 6330 Cham, Switzerland

Contents

About the Editors

Sherif El-Badawy Associate Professor, Mansoura University, Public Works Engineering Department, Faculty of Engineering, Mansoura University, Mansoura, Egypt, PH: +201000183519, E-mail: sbadawy@mans.edu.eg

Sherif El-Badawy is Associate Professor at Mansoura University, Egypt. He serves as the Director of Highway and Airport Engineering Laboratory, and Director of the Center of Scientific, Experimental, and Technical Services, Mansoura University. He received a BSc with honor in Civil Engineering and MSc from Mansoura University, Egypt. Dr. El-Badawy pursued his PhD at Arizona State University (ASU), USA, in 2006. He worked as a Postdoctoral Research Associate at ASU from June 2006 to July 2007 and as a Research Fellow at University of Idaho from December 2009 to January 2012. He has more than 20 years of experience in pavement structural analysis, design, and characterization. He has participated in several state and national projects during his employment in the USA. He was one of the team members under Professor Witczak in the NCHRP 1-37A and 1-40D Projects which focused on the development of the Mechanistic-Empirical Pavement Design Guide (MEPDG). He serves as a TRB Committee Member on Flexible Pavement Design (AFD60) since April 2012. He is an Elected Board Member of the Middle East Society of Asphalt Technologists (MESAT), a Member of the International Geosynthetic Society (IGS), Transportation and Traffic Society (TTS),

International Association of Computer Science and Information Technology (IACSIT), and Chi Epsilon Honor Society. He developed and taught many undergraduate and postgraduate courses in USA and Egypt. His research interest focuses on pavement material characterization and modeling, mechanistic-empirical pavement design methods, and traffic characteristics. He was the Chief Editor of the ASCE GPS "Innovative Technologies for Severe Weathers and Climate Changes." Through research and graduate advising, he published more than 70 technical publications and reports.

Ragaa Abdelhakim Assistant Professor, Tanta University, Public Works Engineering Department, Faculty of Engineering, Tanta University, Tanta, Egypt, PH: +201008258088, E-mail: ragaa.abdelhakim@f-eng.tanta.edu.eg

Ragaa Abdelhakim is an Assistant Professor at Tanta University, Egypt. She received a BSc with honor in Civil Engineering (Structural Engineering) from Tanta University, Egypt. Dr. Abdelhakim pursued her MSc and PhD at Tanta University in Civil Engineering (Public Works Engineering). She worked as an Assistant Professor in Civil Engineering Department, Faculty of Engineering, Delta University for Science and Technology from June 2010 to January 2014. She started working in Tanta University since 2014, and she was a Fulbright Visiting Scholar at Texas A&M University in the academic year 2017/2018. She has taught many undergraduate and postgraduate courses. Her research interest focuses on pavement material characterization and modeling, mechanistic-empirical pavement design methods, and traffic characteristics.

Performance Evaluation of Ultra-Thin Pavement Seals in Low-Volume Roads

Jacob Adedayo Adedeji(✉) and Mohamed Mostafa Hassan

Sustainable Urban Roads and Transportation (SURT) Research Group, Department of Civil Engineering, Central University of Technology, Free State, Bloemfontein, Republic of South Africa
{jadedeji,mmostafa}@cut.ac.za

Abstract. The concept of ultra-thin seals for pavement was initially a preventive maintenance strategy for deteriorating pavement structure. However, it has currently been considered as an alternative for upgrading unsealed low-volume road to sealed road. Considering ultra-thin seal surfacing in low-volume roads will make these roads to be maintenance cost effective, improve ride quality, reduce noise generated, prevent water ingression to underlying layer, and reduce dust generation and loss of aggregate. Nevertheless, the question remains which pavement surfacing option will give a substantial improvement in terms of bearing capacity and economy. Therefore, this paper examines the performance of sealed low-volume roads using asphalt and concrete. Consequently, a finite element (FE) model was developed in general-purpose FE software Abaqus to simulate the pavement layers under a static loading condition and benchmarked with multi-layered elastic software WinJulea and mePADS. In this study two scenarios of low-volume roads were modelled by varying the thickness of the surfacing layer over the conventional granular base and stabilized base layer. Comparative results of the two surfacing options were measured through the horizontal strains underneath the surface layer, vertical strains on the subgrade layer and the increase in pavement bearing capacities. Thus, this study evaluates and suggests the most viable ultra-thin surfacing option between asphalt and concrete for low volume roads.

1 Global Trend of Low Volume Roads

Globally, the reduction in time and cost of travel, and coupled with traffic safety are the ultimate deliverables of a well-designed road system (Brito 2011). However, these are yet to be met as there is a continuous increase in travel time and vehicle operation cost, especially when travelling on Low Volumes Roads (LVRs) which make up more than 80% of the world road network (Ramulu et al. 2012). LVRs provides the major links to the highway transportation system and also the major links between raw materials and markets. Thus, these roads consist a great length of road network throughout the world, for instance, USA, Canada, UK, India and South Africa, has approximately 65%, 75%, 95%, 80% and >75% of LVRs respectively (Hall and Bettis 2000; Doré 2014; Paige-Green 2007; Brito 2011; Gupta 2017; Cook et al. 2013). Nevertheless, the necessities of LVRs to the economic development and alleviating poverty in a country

S. El-Badawy and R. Abd El-Hakim (Eds.): GeoMEast 2018, SUCI, pp. 1–14, 2019.
https://doi.org/10.1007/978-3-030-01908-2_1

cannot be overemphasized, however, the larger percentage of the LVRs around the world remains unsealed and most especially in developing countries.

LVRs definitions vary with the responsible agency and the location, in the USA, it is defined as road which carries less than 500 vehicles per day (vpd) while in South Africa, it is defined to carry an average traffic of less than 500–400 vpd (Hall and Bettis 2000; Ramulu et al. 2012). These roads are classified into 3 categories which are rural roads, recreational and resource development roads (Doré 2014). As result of the limited traffic volume per day based on their functionalities, these roads are not a primary focus of the transportation industry and the government. Therefore, these roads still remain or unsealed and thus, jeopardized the economic growth of the country in terms of the agricultural sector and social services (Jagtap and Nagrale 2013).

Furthermore, 95% of the unsealed roads in South Africa are low volume (Paige-Green 2007). LVRs are considered to carry a traffic less than 200,000 equivalent standard axles (ESAs) over a 20-years design life. In South Africa, these roads are taken to have daily traffic less than 500 vpd and are basically falls under road category C and D, which are descript as “light trafficked rural roads, strategic roads” and “rural access roads” respectively, and regarded as “less important” road. Other characteristics of these roads include; typical pavement class: < ESA0.03 – ESA3 (million standard axles/lane); total equivalent traffic loading (E80/lane): < 3 million depending on design strategy …etc. (SANRAL 2013). Therefore, low volume roads are usually merely graded; just to highlight the travel path or at most addition of graveled surface, without further improvement. Cost cutting is major ground for this practice; however, the cost of operating vehicle on these roads and other aforementioned disadvantages are becoming unbearable as the years go by.

Additionally, unsealed roads have constrained mobility in terms of travel time, significantly increase vehicle operating and maintenance costs, increased accident rates, contribute to dust generation which is hazardous to human’s health, and cause unsustainable loss of gravel material, which is quantified at 30 to 60 million cubic meters annually (Paige-Green 2005; Paige-Green and Pinard 2012). Furthermore, unsealed roads contribute to human and property costs and aggravate isolation, poverty, poor health, and illiteracy in the rural communities. Considering these challenges, a sustainable and economical upgrading of unsealed roads to sealed low volume roads through the ultra-thin seal techniques seem to be the way out. Nevertheless, the question remains which pavement surfacing option will give a substantial improvement in terms of bearing capacity and economy.

1.1 Surfacing Options for Unsealed Low Volume Roads

According to Henning et al. (2006), an unsealed road is a road without permanent waterproof surface (surface layer) which include both engineered and un-engineered roads. An unpaved road in most cases consist of the subgrade layer and proper top-up layer which may serve as the sub-base and in exceptional cases might consist of the base layer. In contrast, sealed roads are all-weather dust-free surface; includes a surface layer, no matter how thin it may be or the type of material. A typical sealed road may or may not consist of all layers (i.e. surface, base, subbase, and subgrade) in a conventional pavement but there will be a seal on the surface. The overall purpose is to prevent water ingression into the underlying layers, improve the riding quality and improve the structural capacity of the pavement structure.

Although, Araya (2011) argues that the main structural element in a pavement is formed by the unbounded granular layer as thick sub-base or/and base layers placed over the subgrade, with a thinning surface layer which majorly provides protection against water ingress (Gupta 2017). Thus, more focus should be given to the base and sub-base layer rather than the surface layer in a typical low volume road, however, to a great extent stabilization of the layers have been considered as it is observed to be yielding great results.

In addition, innovations have brought about different surfacing options which can be considered economical rather than the conventional asphalt or concrete surfacing; however, these are affected by various factors which can further affect their selection for use. Such factors include; climate and topography condition, environmental and socio-economic impact, safety, suitability, the durability of surfacing etc. Surfacing options, the various factors affecting them and selection considerations are presented in Table 1 (Henning et al. 2006; Doré 2014; TRH 4 1996). In addition, selection considerations were based on cost and performance which are further divided as follows; Cost: Initial cost (IC), Maintenance cost (MC), and Rehabilitation cost (RC), and Performances: Service life (SL), Level of service (LoS), and Structural Capacity (SC). It is worthy to note that there are various costs involved in the use of bituminous and concrete surfaces, yet, the performance tends to outshoot these costs and vice versa in the case of gravel surfacing.

1.2 Ultra-Thin Sealing Concept

Overall, the original concept of thin layer surfacing can be referred as a preventive maintenance strategy, which restores pavement back to its original state and can be applied to both flexible and concrete pavements. At first, the ultra-thin surface concept was based on asphalt products only because it main purpose includes reduction of noise level, waterproofing of the underlying layers, and prevention of aggregate loss (Sandberg et al. 2011). Thus, a number of systems were developed for thin asphalt surfacing which includes; thin hot-mix asphalt overlay, open graded friction course, ultra-thin bonded wearing course etc. The major differences in all of these systems are the thickness and this is based on what the surfacing is set to achieve. Using thin surface has increased with the ongoing improvement in their technology and cost-effectiveness (Merritt et al. 2015). Nevertheless, the concept is not limited to asphalt surfacing but alternatively, the use of concrete is now gaining popularity (Pretorius et al. 2004; Groenewald and Van Wijk 2010; Du Plessis et al. 2014). The concrete application of the thin surfacing which is known as ultra-thin reinforced concrete pavement (about 150 mm in thickness) has been applied in both high volume and low volume roads with good success, yet, designers are still faced with a number of uncertainties and there are still questions about the economic and environmental impact when compared with the thin asphalt surfacing option (Du Plessis et al. 2014).

Overall, considering the use of ultra-thin asphalt surfacing in low volume roads will make these roads be cost effective, improve surface texture, reduce noise generated, improve ride qualities, assists waterproofing of underlying surface, and reduce dust

Table 1. Surfacing options for pavement structures and their selection consideration (Henning et al. 2006; Doré 2014)

Surfacing type	Description	Selection consideration[#]					
		Cost			Performance		
		IC	MC	RC	SL	LoS	SC
Natural surfacing	Engineered earth roads or natural surfaces. Generally, have poor geometry and drainage	1	4	1	1	1	1
Gravel	Typically 150–250 mm thick natural gravel or other imported layer that is worn down by traffic and the environment	1	4	2	1	1	1
Dust suppressants	Additionally to a good construction and a mechanical stabilization, dust can be controlled with chemical additives, such as Wetting Agents, Salts/Chlorides, Natural Polymers, Wax Agents, etc.	2	2	3	2	3	1
Stone	Crushed stone layers can be placed with machines or manually. The former requires heavy equipment for compaction. The latter may be prepared without heavy compaction equipment	3	2	3	3	2	3
Bricks	Usually prepared from high-quality clay bricks. Pavements are very durable and can present a very tight, relatively smooth surface	3	2	3	3	3	3
Concrete	Very durable, but mostly require minimum thickness for high volume roads. A special application is the concrete block pavement, with similar behavior and performance to brick and clay bricks	5	2	5	5	4	5
Bituminous surfaces	Classified into two groups: Seals (bitumen film and stone embedded) and Bituminous mixes (asphalt layers)	4	1	4	4	4	4
Other surfaces	Recycled rubble, concrete or asphalt mix	3	2	3	3	3	4

*Where: 5–very high; 4–high; 3–moderate; 2–low; 1–very low

generation and loss of aggregate (Sandberg et al. 2011; Yetkin Yildirim 2015). Nevertheless, the total cost of upgrading an unsealed to sealed road must be kept as low as possible. This can, however, be achieved by; (i) making optimal use of the in-situ materials, thus minimizing the quantity of imported materials and reducing environmental impact (Paige-Green and Pinard 2012), (ii) using simple, cheap and effective material testing methods such as dynamic cone penetrometer (Paige-Green and Pinard 2012; Salgado and Yoon 2003), (iii) considering the use of labor intensive construction practices (TRH 4 1996; Pinard et al. 2015), and (iv) incorporating its design in the current pavement design trend (FEM) for validation and in order to estimate the possible structural maintenance/renovation (Du Plessis et al. 2014). Largely, based on the scope of this study the aspect of the designing of the low volume roads using the current design trend will be considered.

2 Finite Element Modeling of Low Volume Roads

Using mechanistic-empirical design methods ME-DM over traditional methods is continually gaining interest over decades (Kim 2007). This results from the inaccuracy of design and failure experienced in the use of traditional methods (Huang 2004; Jain et al. 2013), and the advantages of using ME-DMs, are the ability to determine stress and strain, to analyze static and time-dependent problems, to incorporate non-linear material characterization, large strain/deformations, dynamics analysis and other sophisticated features. The ME-DM analysis is based on the inputs such as pavement thickness, material properties, loading condition (in terms of static and dynamic) and other substances interacting with it. Thus, to model pavements correctly, it is necessary to use numerical methods, such as the finite difference method, the boundary element method and FEM (Áurea et al. 2006; Ameri et al. 2012). However, FEM is the most adopted in pavement analysis and thus, will be considered.

Essentially, various research studies have been undertaken on the layers in flexible pavement via FEM (Gupta and Kumar 2014). Yet granular materials do not feature strongly, as more focus is given to designing the asphalt layer and subgrade condition (Adu-Osei 2001; Araya 2011; Tiliouine and Sandjak 2014). Similarly, only limited work has been undertaken on stabilized base and subbase layers (Peng and He 2009). Nevertheless, there are various sources of error in pavement performance predictions and most are more difficult to control than the response model (NCHRP 2004). Therefore, a reality check through validation of results with field testing is of importance.

However, FEM has been successfully used in the analysis of the major forms of failure in pavement structure such as rutting and fatigue cracking at different layers (Walubita and van de Ven 2000; Al-Khateeb et al. 2011; Abed and Al-Azzawi 2012), and also used to determine the accurate positioning of geogrid materials (Al-Azzawi 2012), thickness of each layer (Shafabakhsh et al. 2013; Sinha et al. 2014) and the interaction between pavement and its instrumentation (Zafar et al. 2005; Yin 2013). Recently, researchers have started considering the use of FEM for the design of low volume roads. Jagtap and Nagrale (2013), used FEM for design of low volume road and the study shows that it reduces construction cost by 11–15%. Considering all these successes, it is also important to feature the design of low volume roads in FEM for a perpetual pavement structure.

3 Failure Models in FEM

Distress prediction models are derived from observation and performance of pavement with relations to observed failure and initial strain under various loads, thereby computing the number of loading cycles to failure (pavement life span) (Mamlouk and Mobasher 2004; NCHRP 2004; Pavement Interactive 2008). Various distress prediction models exist, nonetheless, two are widely recognized; fatigue cracking in asphalt and deformation in the subgrade (Pavement Interaction 2008; Ekwulo and Eme 2009; South African National Road Agency Ltd. (SANRAL) 2013). Furthermore, Asphalt Institute damage model (Asphalt Institute 1982) is commonly accepted (Pavement Interaction 2008; Ekwulo and Eme 2009). Asphalt Institute models are presented in Eq. 1 (Fatigue criterion) and 2 (Rutting criterion).

$$\mathrm{N_f} = 0.0796(\varepsilon_\mathrm{t})^{-3.291}(\mathrm{E})^{-0.854} \tag{1}$$

Where; N_f = Number of repetitions for fatigue cracking; ε_t = Tensile strain at the bottom of the asphalt surface in microstrain; E = resilient modulus of asphalt in psi.

$$\mathrm{N_r} = 1.365 \times 10^{-9}(\mathrm{E_c})^{-4.477} \tag{2}$$

Where; N_r = Number of repetitions for subgrade rutting failure; E_c = Compressive strain on top of the subgrade.

$$N_{eff} = SF \times 10^{c\left(1-\frac{\varepsilon}{d\varepsilon_b}\right)} \tag{3}$$

Where; N_{eff} = Effective fatigue life; SF = Shift Factor for crack propagation; ε = Horizontal tensile strain at bottom of layer in microstrain; ε_b = Strain-at-break; c, d = Constants (SANRAL 2013). Note assumptions for rigid pavement is taken from SANRAL (2013).

Additionally, in the rigid pavement, the effective fatigue life is used in the estimation of the lifespan of the cemented or concrete layer to failure. The horizontal tensile strain at the bottom of the layer is used in the Eq. 3 with other constant which was established (SANRAL 2013). Overall, the distress prediction models are used to define the point at which failure occurs in a pavement by determining the incremental damage.

4 Design Model for Ultra-Thin Seals in Low Volume Roads

In this study, a typical low volume road was modelled. A three layered pavement structure was considered and consist of 20–40 mm surface options (Asphalt/Concrete), 300 mm stabilized base/granular base and 2000 mm subgrade layer. Four models were modelled per pavement surfacing options (Fig. 1). These models highlight the effect of thickness of surfacing options and the type of pavement surfacing. Similarly, all models were modelled in WinJulea for validation and comparative study.

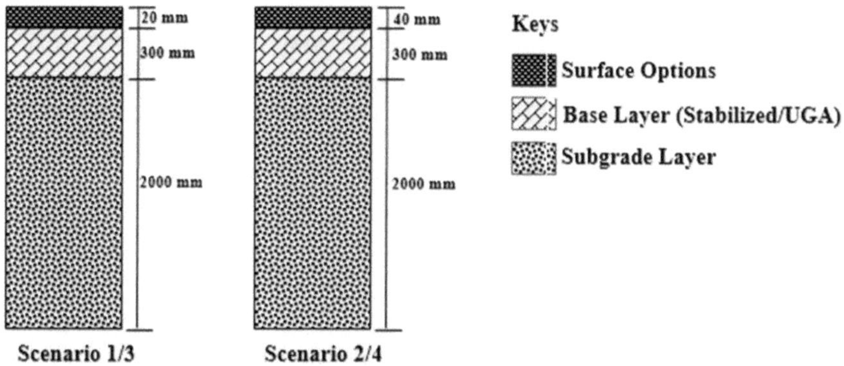

Fig. 1. Model scenarios of pavement structure per pavement surfacing options

4.1 Model Geometry

The dimensions of the simulated model are 3000 mm length by 3000 mm breath and the total depth varies based on the thickness of the surface layer as the case may be. This geometry is similar to that used by Ahmed (2006), with the aim of avoiding edge error when loaded. However, the depth of the base layer and the subgrade layer remain constant at 300 mm and 2000 mm respectively. A solid continuum 8-noded element with reduction integration (C3D8R) was used in the element type and the meshing around the loading area are fine and tends to coarse at distances away from the applied load for an efficient model (Tiliouine and Sandjak 2014). This was achieved using the Bias local seeding methods. Thus, the model was seeded at 2.5 mm at the loading area and 50 mm–500 mm away from the loading area. Additionally, in WinJulea and MePADS, the flexible pavement is model as an axisymmetric structure subjected to a circular load with linear elastic materials. Furthermore, WinJulea and MePADS were used as a control and also to determine the mesh configuration of the finite element model.

4.2 Material Characterization

All layers of the pavement are assumed to be linearly elastic in behavior for simplicity as non-linear properties require many input parameters which are not readily available (Al-Jhayyish 2014). Additionally, material characterization for asphalt, gravel and subgrade are selected from SANRAL (2013) and fly ash stabilized material is adapted from a previous study by Adedeji (2015), these material characterizations are presented in Table 2. The material characterizations are present in both International System of Units and US customary units to accommodate for the software (Abaqus and WinJulea) used in this research.

Table 2. Material characterizations used in models

Pavement layer	Material code (TRH 4 1996)	Layer thickness (mm)	inch	Elastic modulus (MPa)	psi	Poisson ratio
Asphalt	AG	20	**0.79**	3000	**580151**	0.40
Concrete	–	20	**0.79**	30000	**4351132.1**	0.2
Stabilized gravel	C4	300	**11.8**	2560	**371297**	0.35
Gravel	G4	300	**11.8**	300	**43511**	0.35
Subgrade	G7	2000	**78.74**	120	**17405**	0.35

4.3 Model Interactions and Conditioning

On boundary and loading condition aspects, the models are all subjected to a static load in a linear perturbation analysis and the models are restrained in horizontal directions (i.e. the degree of freedom 1 and 3) with the subgrade base in all directions. Further, loading contact area is assumed to be rectangular (119355 mm^2) (i.e. 285 $\times$ 419 mm) and Abaqus Axisymmetric model is assumed to be radius of a circle (195 mm) with wheel pressure of 0.67 Mpa. Additionally, tie constraints are assumed as the interaction between the interfaces of the layers (i.e. layers are fully bounded with no friction) for control. On the other hand, in WinJulea and MePADS, the load contact area is taken as 185 inch^2 (119355 mm^2 as in Abaqus model) and 195 mm respectively with a load of 17985 lbs (80000 N).

5 Results of FEM Analysis and Discussion

This study analyse the performance of ultra-thin pavement seals using the asphalt and the concrete surfacing option for low volume roads. Against the customary standards of 50 mm thickness surfacing option for pavement structure, this study considered the use of thinner surfacing option and compares the most viable between rigid and flexible pavement. This is, however, to put the cost of low volume road as low as possible without compromising the level of service and structural capacity. In order to measure the effect of the surface layer, critical areas such as the surface deflection, horizontal strain at the bottom of the surface layer and the cemented layers, compressive vertical strain/stress at the top of the subgrade, are used for the pavement verification analysis.

5.1 Validation of FEM

Validation of the FEM was done to check the results (in terms of stress and strains) given by it with linear elastic theory results (Gupta 2017). Scenario 1 using Asphalt layer of 20 mm, Stabilized layer 300 mm and Subgrade 2000 mm, and linear elastic properties for all layers were used for the validation analysis. WinJulea and MePADS, linear elastic layered analysis programs were used to validate the FEM. Furthermore, Axisymmetric model in Abaqus was modelled and results give the very close result to those of layered elastic software Fig. 2. Details of the design model used in the WinJulea and MePADS have been highlighted above. Additionally, the results of the 3D FEM model differ from the other by 30% for the vertical strain on the subgrade and lesser for other layers despite similar mesh configuration was used (Table 3).

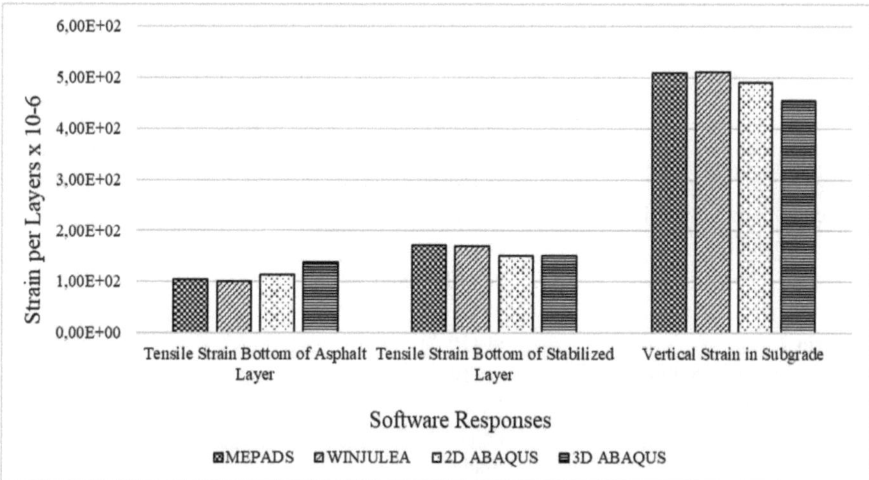

Fig. 2. Strain values for pavement layers in MePADS, WinJulea, 2D Abaqus and 3D Abaqus model

5.2 Effect of Surfacing Thickness on Pavement Behavior

Increase in pavement surfacing thickness should decrease the compressive strain transferred to the subgrade layer, however, this is a factor of the type of surfacing option and the type of base material used. Comparing scenario 1 (Surface 20 mm + Stabilized Base + Subgrade) and 2 (Surface 40 mm + Stabilized Base + Subgrade) for Asphalt pavement, it is worth noting that the horizontal strain at the bottom of the surface and stabilized base layer and the vertical strain on the subgrade decreased from (138.6×10^6 to 117.7×10^6), (149×10^6 to 133.1×10^6) and 455×106 to 405.2×10^6 respectively using the 3D FEM model. Similarly, this is the same case in Scenario 3 and 4 using Gravel as the base layer (Table 4).

In rigid pavement model, it was noted that the increase in surface layer decreases the horizontal strain at the bottom of the surface layer and the compressive strain at the

Table 3. Model mesh configuration for Abaqus axisymmetric and 3D FEM model

Pavement structures	Scenario 1 Axisymmetric model	Scenario 1 3D FEM model	Scenario 1 Axisymmetric model	Scenario 1 3D FEM model
	No. of elements along the layer thickness	No. of elements along the layer thickness	Total number of elements CAX4R	Total number of elements C3D8R
Surface layer	4	4	256	45212
Stabilized gravel	16	16	1024	361696
Subgrade	6	6	384	113030

Table 4. Pavement responses in asphalt versus rigid pavement

	Pavement responses	Asphalt pavement	Rigid pavement
Scenario 1 (Surface 20 mm + Stabilized Base + Subgrade)	**εt (10^{-6})** Surface	138.6	47.9
	εt (10^{-6}) Stabilized Base	149.0	25.3
	εc (10^{-6}) Sub-grade	455.0	371.8
Scenario 2 (Surface 40 mm + Stabilized Base + Subgrade)	**εt (10^{-6})** Surface	117.7	22.4
	εt (10^{-6}) Stabilized Base	133.1	111.0
	εc (10^{-6}) Sub-grade	405.2	318.6
Scenario 3 (Surface 20 mm + Gravel + Subgrade)	**εt (10^{-6})** Surface	174.0	74.16
	εt (10^{-6}) Stabilized Base	–	–
	εc (10^{-6}) Sub-grade	1266.0	1141.0
Scenario 4 (Surface 40 mm + Gravel + Subgrade)	**εt (10^{-6})** Surface	151.3	58.29
	εt (10^{-6}) Stabilized Base	–	–
	εc (10^{-6}) Sub-grade	1123.0	864.8

top of the subgrade layer as seen in the flexible pavement. Furthermore, the horizontal strain at the bottom of the stabilized layer in rigid pavement model (scenario 1 and 2) increases rather than decreasing as in the case of asphalt pavement; thus, implies that the stabilized base layer with rigid surface increase will quickly experience crushing and then fail eventually.

5.3 Effect of Ultra-Thin Surface Type on Pavement Life Span

Using Eqs. 1, 2 and 3, the pavement life spans were calculated for the fatigue and rutting criterion as shown in Table 5. The results show that the effective fatigue for concrete surface thickness less than 102 mm will give a proximity same results and this is justified on the concept that ultra-thin concrete surfacing option is usually taken as a minimum of 150 mm. Although, the rigid pavement under the fatigue criterion perform better than the flexible pavement except for scenario 2 where the flexible pavement performs better. Thus, in the design of ultra-thin pavement seal, the use of 40 mm thickness of asphalt layer will perform well and it is also worthy to note that the stabilized layer in the rigid pavement for scenario 2 will also fail first. Overall, the use of concrete surface layer is expensive from the initial cost and less expensive with the maintenance cost when compared with the flexible pavement. If the cost in term of total cost for both pavement structures are kept constant, the flexible pavement performance will out shoot that of rigid pavement. Nevertheless, the use of waste and recycling of materials should be encouraged in the construction of low volume road so as to attain the sustainability target.

Table 5. Fatigue and rutting failure analysis for asphalt and rigid pavement

	Fatigue criterion No. of load Repetitions to Failure N_f/N_{eff}		Rutting criterion No. of load Repetitions to Failure N_r	
	Asphalt pavement	Rigid pavement	Asphalt pavement	Rigid pavement
Scenario 1	4.8×10^6	7.4×10^6	1.3×10^6	3.1×10^6
Scenario 2	8.1×10^6	7.4×10^6	2.1×10^6	6.2×10^6
Scenario 3	2.3×10^6	7.4×10^6	12.8×10^3	20.6×10^3
Scenario 4	3.6×10^6	7.4×10^6	21.9×10^3	70.5×10^3

6 Summary and Conclusion

In the present study, four scenarios models have been implemented in a 3D FEM Abaqus to understand the performance of ultra-thin pavement seals such as flexible and rigid pavement in low volume roads. The 3D FEM Abaqus model was benchmarked by multilayered elastic software (WinJulea, MePADS) and 2D FEM Abaqus. The validation results show good agreement between WinJulea, MePADS and 2D FEM model and the 3D FEM model differ from the others by 30% for the vertical strain on the subgrade and lesser for other layers despite similar mesh configuration to 2D FEM model was used. Overall, the use of ultra-thin surfacing option in low volume roads cannot be underestimated. Although, on a school of thought says the use of surface layer would only enhance riding quality, prevent water ingression, and reduce the dust generation on low volume roads, yet, it is worthy to note that the ultra-thin seal surfacing option would strengthen the pavement structural capacity. However, the question answered in this paper is which pavement surfacing option will give a substantial improvement in terms of bearing capacity and economy? Results show that an asphalt surfacing option at a thickness of 40 mm will perform better compared to a 40 mm concrete pavement. Although, the subgrade will fail first, however, the riding quality will still be better than that of rigid pavement. Further studies can be carried out considering the nonlinearity effect of the gravel base layer on the surfacing options for low volume roads.

References

Abed, A.H., Al-Azzawi, A.A.: Evaluation of rutting depth in flexible pavements by using finite element analysis and local empirical model. Am. J. Eng. Appl. Sci. **5**(2), 163–169 (2012)

Adedeji, J.A.: Simulation of flexible pavement utilizing fly ash as alternative stabilizer. Master's thesis, Department of Civil Engineering, Central University of Technology, Free State, South Africa (2015)

Adu-Osei, A.: Characterization of unbound granular layers in flexible pavements. Technical reported, Texas A&M University Texas Transportation Institute College Station, Texas (2001)

Ahmed, H.Y.: Effect of surface cracking on responses of flexible pavements structure. J. Eng. Sci. **34**(3), 699–717 (2006)

Al-Azzawi, A.A.: Finite element analysis of flexible pavements strengthened with geogrid. ARPN J. Eng. Appl. Sci. **7**(10), 1295–1299 (2012)

Al-Jhayyish, A.K.: Incorporating chemical stabilization of the subgrade in pavement design and construction practices. Ph.D. thesis, Ohio University (2014)

Al-Khateeb, L.A., Saoud, A., Al-Msouti, M.F.: Rutting prediction of flexible pavements using finite element modelling. Jordan J. Civ. Eng. **5**(2), 173–189 (2011)

Ameri, M., Salehabadi, E.G., Nejad, F.M., Rostami, T.: Assessment of analytical techniques of flexible pavements by final element method and theory of multi-layer system. J. Basic Appl. Sci. Res. **2**(11), 11743–11748 (2012)

Araya, A.A.: Characterization of unbound granular materials for pavement. Master's thesis, Road and Railway Engineering Section, Faculty of Civil Engineering and Geosciences, Delft University of Technology (2011)

Asphalt Institute: Research and Development of Asphalt Institute's Thickness Design Manual. 9th edn., Research Report 82-2, the Asphalt Institute (1982)

Áurea, S.H., Evandro, P.J., Lucas, T.B.: Finite element modelling of flexible pavement (2006)

Brito, L.: Design methods for low volume roads. Ph.D. thesis, University of Nottingham (2011)

Cook, J.R., Petts, R.C., Rolt, J.: Low Volume Rural Road Surfacing and Pavements, A Guide to Good Practice. OTB Engineering UK LLP, London (2013)

Doré, G.: Pavement design for low volume roads CTEP – APWA Pavement Workshop 2014 – Presentations (2014). http://ctep.ca/wp-content/uploads/2017/05/Pavement-Design-for-Low-volume-Roads-Dore.pdf

Du Plessis, L., Kilian, A., Mngaza, K.: Ultra-thin reinforced concrete pavements (UTRCP): addressing the design issues. In: Proceedings of the 33rd Southern African Transport Conference (SATC 2014), Pretoria, South Africa, pp. 179–190 (2014)

Ekwulo, E.O., Eme, D.B.: Fatigue and rutting strain analysis of flexible pavements designed using CBR methods. Afr. J. Environ. Sci. Technol. **3**(12), 412–421 (2009)

Groenewald, M., Van Wijk, I.: Ultra-thin reinforced concrete pavements (UTRCP)-innovative technology which offers cost and socio-economic benefits for infrastructure provision. SATC 2010 (2010)

Gupta, A., Kumar, A.: Comparative structural analysis of flexible pavements using finite element method. Int. J. Pavement Eng. Asph. Technol. (PEAT) **15**(1), 11–19 (2014)

Gupta, A.: Finite element analysis of granular pavements considering material nonlinearity. Asian Transp. Stud. **4**(3), 550–564 (2017)

Hall, K.D., Bettis, J.B.: Development of comprehensive low-volume pavement design procedures. MBTC 1070, Final Report (2000)

Henning, T., Kadar, P., Bennett, C.R.: Surfacing alternatives for the unsealed rural road. The world bank, Washington, DC Transport note no. Trn-33 (2006)

Huang, Y.H.: Pavement Analysis and Design, 2nd edn. Pearson Education Inc., Hoboken (2004)

Jagtap, P.S., Nagrale, P.P.: Benefits of mechanistic approach for low volume rural roads. In: Proceedings of Indian Geotechnical Conference, Roorkee, pp. 1–8 (2013)

Jain, S., Joshi, Y.P., Golia, S.S.: Design of rigid and flexible pavements by various methods and their cost analysis of each method. Int. J. Eng. Res. Appl. **3**(5), 119–123 (2013)

Kim, M.: Three-dimensional finite element analysis of flexible pavements considering nonlinear pavement foundation behaviour. Department of Civil Engineering, Graduate College of the University of Illinois at Urbana-Champaign (2007)

Mamlouk, M., Mobasher, B.: Cracking resistance of asphalt rubber mix versus hot-mix Asphalt. Int. J. Road Mater. Pavement Des. **5**(4), 435–452 (2004)

Merritt, D.K., Lyon, C.A., Persaud, B.N.: Evaluation of Pavement Safety Performance (No. FHWA-HRT-14-065) (2015)

NCHRP: Guide for Mechanistic-Empirical Design of New and Rehabilitated Pavement Structures: Appendix RR-Finite Element Procedures for Flexible Pavement Analysis, Transportation Research Board, National Research Council, Washington, D.C. (2004)

Paige-Green, P.: Practical aspects of low cost sealing of roads (2005). http://www.ssatp.org/sites/ssatp/files/publications/HTML/LVSR/English/Miscellaneous/02-Paige-Green-CSIR-LowCostSeals-2005.pdf. Accessed 5 Jan 2016

Paige-Green, P.: Local government note: new perspectives of unsealed roads in South Africa. In: Keynote Address Presented at the REAAA (NZ) Low Volume Roads Workshop held in Nelson, New Zealand, 18–20 July 2007 (2007)

Paige-Green, P., Pinard, M.I.: Optimum design of sustainable sealed low volume roads using the dynamic cone penetrometer (DCP). In: 25th ARRB Conference—Shaping the Future: Linking Policy, Research and Outcomes, Perth, Australia (2012)

Pavement Interactive: Flexible pavement mechanistic models (2008). www.pavement-interactive.org/article/flexible-pavement-mechanistic-models/. Accessed 27 Mar 2014

Peng, Y., He, Y.: Structural characteristic of cement-stabilized soil bases with 3D finite element method. Front. Archit. Civ. Eng. China **3**(4), 428–434 (2009)

Pinard, M.I., Paige-Green, P., Hongve, J.: Developments in low volume roads technology: challenging conventional paradigms. In: 11th Conference on Asphalt Pavements for Southern Africa (CAPSA15), Sun City, South Africa (2015)

Pretorius, F.J., Wise, J.C., Henderson, M.: Development of application differentiated ultra-thin asphalt friction courses for southern African application. In: Proceedings of the 8th Conference on Asphalt Pavements for Southern Africa (CAPSA 2004), vol. 12, p. 16 (2004)

Ramulu, G., Shankar, S., Chowdary, V., Prasad, C.S.R.K.: Influence of unbound material properties on rutting potential of low volume roads. Elixir Cem. Concr. Compos. **42**, 6377–6382 (2012)

Salgado, R., Yoon, S.: Dynamic cone penetration test (DCPT) for subgrade assessment. Joint Transp. Res. Progr. **73** (2003)

Sandberg, U., Kragh, J., Goubert, L., Bendtsen, H., Bergiers, A., Biligiri, K.P., Karlsson, R., Nielsen, E., Olesen, E., Vansteenkiste, S.: Optimization of thin asphalt layers. In: ERA-NET ROAD Project, Swedish National Road and Transport Research Institute (VTI), Danish Road Institute (DRI) and Belgian Road Research Centre (BRRC), vol. 30 (2011)

Shafabakhsh, G.A., Motamedi, M., Family, A.: Influence of asphalt concrete thickness on settlement of flexible pavements. Electron. J. Geotech. Eng. **18**, 473–483 (2013)

Sinha, A.K., Chandra, S., Kumar, P.: Finite element analysis of flexible pavement with different subbase materials. Technical paper. Indian Highw. **42**(2), 53–63 (2014)

South African National Road Agency Ltd (SANRAL): Pavement Design: South African Pavement Engineering Manual. An initiative of the South African National Roads Agency Ltd., South Africa (2013)

Tiliouine, B., Sandjak, K.: Numerical simulation of granular materials behavior for unbound base layers used in Algerian pavement structures. Int. J. Civ. Struct. Eng. **4**(3), 419–429 (2014)

TRH 4: Structural design of flexible pavements for interurban and rural roads. Pretoria: Committee of Land Transport Officials (COLTO), Department of Transport (DoT). DoT Technical Recommendations for Highways; Draft TRH 4 (1996)

Walubita, L.F., van de Ven, M.F.: Stresses and strains in asphalt-surfacing pavements. In: South African Transport Conference Organised by: Conference Planners, 'Action in Transport for the New Millennium' (2000)

Yetkin Yildirim, P.: Texas Considers Ultra-Thin HMA Alternatives to Seal Coats - TSP2 Pavement Preservation (2015). Tsp2pavement.pavementpreservation.org. https://tsp2pavement.pavementpreservation.org/2015/03/30/texas-considers-ultra-thin-hma-alternatives-to-seal-coats/. Accessed 30 July 2016

Yin, H.: The impact of strain gage instrumentation on localized strain responses in asphalt concrete pavements. Int. J. Pavement Res. Technol. **6**(3), 225–234 (2013)

Zafar, R., Nassar, W., Elbella, A.: Interaction between pavement instrumentation and hot-mix-asphalt in flexible pavements. Emir. J. Eng. Res. **10**(1), 49–55 (2005)

Reuse of Oil Drilling Cuts for Road Construction in the Eastern Region of Colombia

Guillermo Soto and Gloria Beltran(✉)

Universidad Nacional de Colombia, Bogotá, Colombia
{gasotoc,gibeltranc}@unal.edu.co

Abstract. The search for new oil and gas deposits worldwide has increased due to the high demand for energy sources needed by industrial development and population growth. Although the industry has developed many processes and technologies to minimize waste generation and to dispose them safely and economically, a significant volume of solid waste from oil and gas drilling is expected in the future. For that reason, it is necessary to study some waste reuse alternatives.

In this work, the potential use of ashes derived from oil drilling cuts (ODC), applied to tertiary roads construction for remote areas in Colombia, is evaluated. To that purpose, ashes exposed to mechanical heat treatment are used as an improvement agent for lateritic clayey soils of low plasticity, predominant in the eastern region of Colombia (extended to Brasil and Venezuela). This region was selected due to its complexity: it has limited resources and large difficulties to obtain suitable materials for road construction, it is exposed to extreme cycles of drought and flood with an 8-month rainy season, and this is one of the areas to be recovered after peace agreements in the country with the guerrilla.

In the experimental plan, basic characterization tests were conducted to the soil, the ash, and mixtures soil-ash, as well as strength and durability evaluation of compacted and cured samples of stabilized soil at 0, 7 and 14 days, considering typical climate conditions of temperature and humidity in the region. Based on successful previous experiences, combination with other nonconventional soil stabilizers, such as salts and waterproofing materials, are also been considered to improve some properties of ash cuttings - soil mixtures. As a result, an economical and viable option is obtained to reduce environmental impacts of wastes and to improve lateritic soils of low mechanical characteristics for road earthworks.

Keywords: Waste reuse · Oil drill cuttings · Soil improvement
Lateritic soils

1 Introduction

The huge eastern region of Colombia has practically no proper road infrastructure, due to its geographical and geomorphological characteristics: floodplains, the prevailing climate conditions (cycles of drought and flooding with a rainy season that lasts for

S. El-Badawy and R. Abd El-Hakim (Eds.): GeoMEast 2018, SUCI, pp. 15–23, 2019.
https://doi.org/10.1007/978-3-030-01908-2_2

8 months), the existing subgrade soils (clayey lateritic soils whose behavior varies with water content) and the lack of sources of materials suitable for road construction that increase the material carrying costs from far away. Therefore, any consideration that seeks to improve local soils, with natural or artificial additives easy to use and available in the region, would contribute to solve the problem.

Otherwise, this area is also the most extensive in oil exploration and exploitation wells in Colombia and produces a large amount of solid ash waste from Thermal Desorption Treatment (TDT) of oil drilling cuts (ODC). This waste has become an environmental and economic problem for the oil industry, due to the huge volume and the costs derived from its treatment and final disposal.

In this work, an alternative is addressed to deal with both, the lack of proper materials for road construction and with waste management, to provide a technically, environmentally and economically viable solution. Thus, solid ash waste is reused as a potential soil improvement additive for road construction in the study area.

Although this solid waste has been successfully used in some parts of the world for construction applications, such as concrete blocks (Meegoda and Muller 1993), granular sub-bases (Tuncan and Koyuncu 2000), and hot asphalt mixtures (Meegoda and Muller 1993), its use as an additive for lateritic soils improvement for road embankments or earthworks is still unexplored.

Previous investigations with lateritic soils of the eastern region of Colombia showed good resistance and rigidity responses to saline stabilizers, but they are not entirely satisfying in terms of durability (Caballero 2017). In the present investigation it was demonstrated that the ODC contributes to hydrophobicity and therefore to durability of soils stabilized with sodium silicate. Samples of stabilized mixtures were prepared and tested at wet-drying cycles; climate conditions of the region were also simulated for their curing process.

2 Methodology

2.1 Site Investigation, On-Site Testing and Sampling

The field work phase was defined according to the type and quantity of tests required for soil characterization purposes. Five excavations were done and Dynamic Cone Penetration Tests (DCP) were performed to determine the subgrade capacity. Enough disturbed and undisturbed samples were recovered from the excavations for laboratory testing.

As for the solid ash waste from Thermal Desorption Treatment (TDT) of oil drilling cuts (ODC), they were supplied by an oil exploration company in the region. This material has the appearance of ash of silty sandy texture and has difficulties for characterization purposes. The fine fraction does not have plasticity. It comes from mechanical crushing of sedimentary rock formations mixed with oil-based drilling muds.

2.2 Basic Material Characterization

Tests for characterization of both, ash and collected soils of the study area, were carried on for evaluating physical properties: liquid and plastic limit, activity, water content, specific gravity, unit weight, bearing capacity.

Subsequently, Optimal Compaction Conditions (Optimum Water Content) for Soil and Soil-Ash Were Determined by Varying the Amount of Ash Mixed to the Soil.

Additionally, it was considered an alternative of combined improvement, to take advantage of the good results achieved in terms of strength and stiffness, based on the documented experiences on the use of sodium silicate as an additional agent (Caballero 2017). For this case, diluted sodium silicate in the compaction water were added to new soil-ash mixtures samples, for evaluating the effects on strength, stiffness and durability as well as significant microstructural changes. In summary, the following options were evaluated: Soil (S), soil-ash (SA), soil-silicate (SS) and soil-silicate-ash (SSA).

2.3 Samples Compaction Process

Soil-ash mixtures samples were prepared under optimal conditions and compacted under static load, to be used in physical characterization, resistance and durability tests.

A Versatester machine was used for all the compacted samples required, in layers of 2.0 cm thickness, as shown in Fig. 1. For this purpose, two metal jackets of 3.5 cm and 5.0 cm in diameter and 10 cm in height were used.

Fig. 1. Sample compaction process in laboratory

2.4 Sample Curing Process

In many stabilized soils, the curing process affects the results of the improvement; hence the expected field conditions should be reproduced as realistically as possible in the laboratory. In this work, it was possible to cure the samples under the typical conditions of relative humidity and temperature of the study area; this process was carried out in a stability chamber simulating a temperature of 30° and a relative humidity of 70%, representative of the dry season, when construction works are expected. The curing process was carried out for three (3) ages corresponding to zero

(0), seven (7) and fourteen (14) days to assess the incidence of curing age on resistance and durability.

2.5 Strength Tests

Resistance characteristics of natural soil and soil mixtures with the improvement agents (silicate and ash) were evaluated. Unconfined compressive strength - UCS tests were carried out according to ASTM 2166, using a Universal Shimadzu - AGX machine, at 1 mm/min controlled deformation speed. In Fig. 2, images of specimens under unconfined compression test are presented; from left to right: Soil (S), Soil Ash (SA), Soil Silicate (SS) and Soil Silicate Ash (SSA). Additionally, direct shear tests were performed to determine cohesive and frictional resistance parameters, useful for earthworks designing purposes.

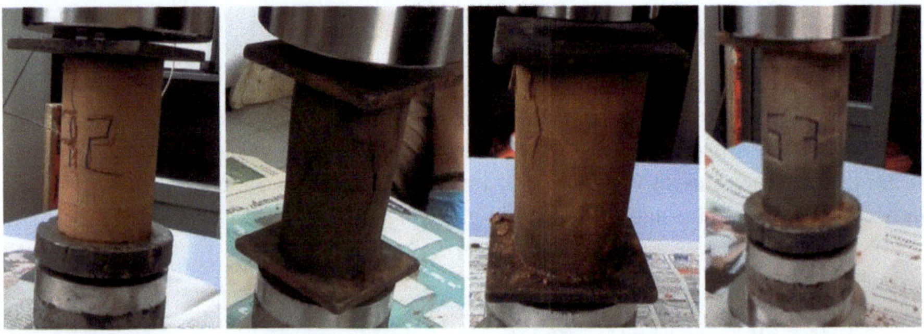

Fig. 2. Samples tested under UCS test

2.6 Durability Tests

The term "durability" refers to the resistance of a material to losing its physical properties or appearance, because of weathering process. To evaluate changes in physical appearance and resistance, two tests were adapted to stabilized soils under: Slake - durability and wetting - drying tests. These tests were conducted under controlled conditions in the laboratory to simulate the effect of weathering process expected in situ.

2.6.1 Slake Durability Test

Originally, the purpose of this test is to determine the resistance of a soft rock sample to softening and disintegration when it is exposed to two cycles of wetting-drying patterns, taking as reference the ASTM D4644 standard. The test consists of inducing the material deterioration during two cycles under saturated condition, while the sample rotates slowly under the axis of the drum of the equipment of slake durability for 10 min in each cycle.

In our case, some specimens showed high values of compressive strength, even comparable with some soft rock. Therefore, it was decided to apply this test to the improved soils.

2.6.2 Wetting Drying Cycles

In order to simulate the effect of drought and flood weather conditions prevailing in the study area, on properties and behavior of different mixtures of soil, five cycles of wetting and drying were applied. Then, unconfined compression tests were conducted after 1, 3 and 5 complete wetting and drying cycles. Although there is no standard for this test, the ASTM D559 provides a guide to the process to be performed, which consist of a wetting time of 5 h and a drying process of 42 h at a constant temperature of 70 °C.

3 Tests Results

The comprehensive analysis of results obtained from all the tests performed, allows to make comparisons, to observe relationships and to appreciate the incidence of: additives, curing time and proportions used, on the strength and durability response of the stabilized soils.

Low plasticity clayey soils (CL) are predominant in the study area, whose natural water content is close to 12.4% and liquid limit and plastic limit of 30.7% and 17.5% respectively. The activity in methylene blue tests was low.

The Dynamic Cone Penetration Tests (DCP) results showed an average Penetration Index of 60 mm/blow, implying a very poor bearing capacity of the natural subgrade in the study area. Consistently, the California Bearing ratio (CBR) test results on undisturbed samples showed indexes of 5%, which decays to 2% after immersing the samples in water.

The optimum water content obtained for natural soils varied between 9% and 12% and the average dry unit weight was 1770 kg/m^3. For soils added with ash, the unit weight increased to 2050 kg/m^3 and the optimum water content to 14%. Finally, soil-silicate improvement test samples showed an optimum water content of 15% and a unit weight of 1800 g/cm^3.

It was determined, by means of multiple tests, that the optimum mixture was a combination of 50% sodium silicate dosed into the compaction water of the sample and 20% of ash dosed by weight. Higher ash proportions appear to be unfavorable in the immediate unconfined compression strength (UCS) (See Fig. 3).

As expected, the curing process allows to develop the resistance with time properly, as shown in Fig. 4, where contrast of the unconfined compression strength (UCS) values is observed; soil- silicate-ash samples showed increases of 200% in resistance compared to natural soil.

Similarly, it was observed in laboratory that the addition of ash contributes to the hydrophobic properties development. Furthermore, by adding the silicate, not only the unconfined compressive strength is improved, but also the durability in samples exposed to wetting-drying cycles, as shown Fig. 5.

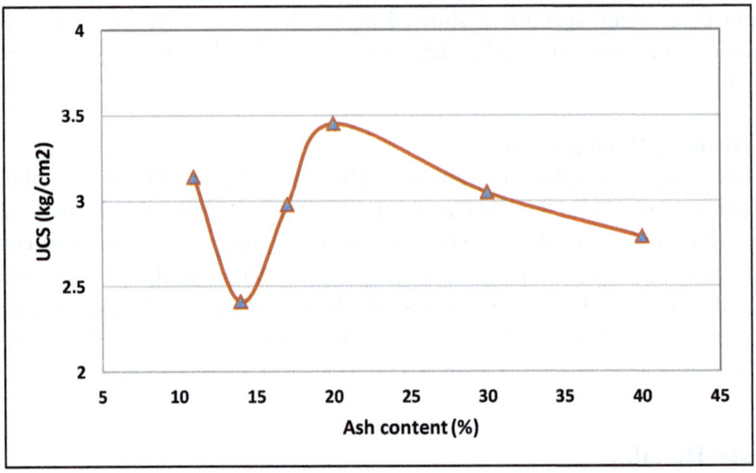

Fig. 3. Strength variation of uncured soil samples with ash content.

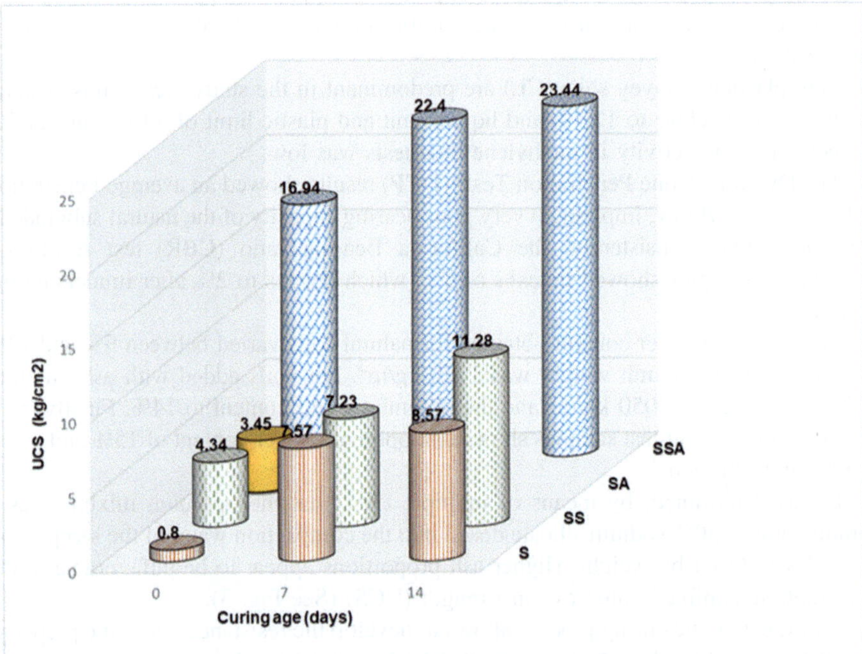

Fig. 4. Summary of the UCS test results

As to slake durability test results, soil samples with no improvement, immediately crumble at the beginning of the first cycle regardless of the curing time; In fact, this behavior represents the actual conditions of soils in the study area during the rainy season. In contrast, samples of soil – ash (SA), soil- silicate (SS) and soil – silicate - ash (SSA) withstand the two test cycles.

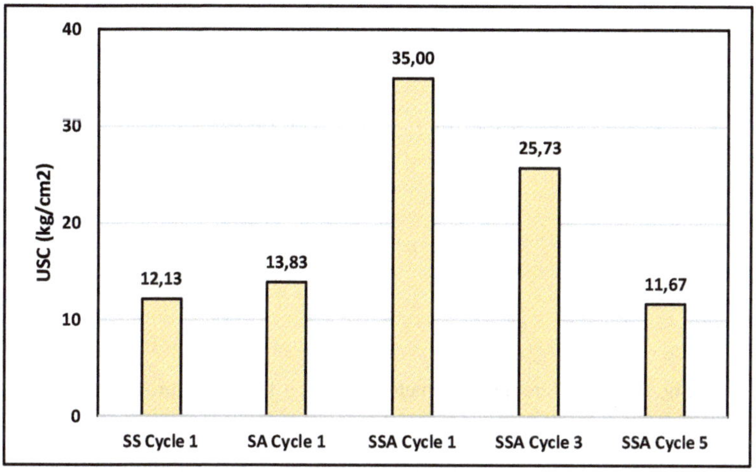

Fig. 5. Results of UCS tests in samples exposed to wetting-drying cycles

In Fig. 6, a bar chart is presented to show variations of the "Slake – Durability Index" (SDI) after two test cycles. The combined soil-ash-silicate (SSA) improvement seems to have the best performance, achieving SDIs from 89% for uncured samples and around 99% for those cured at 14 days.

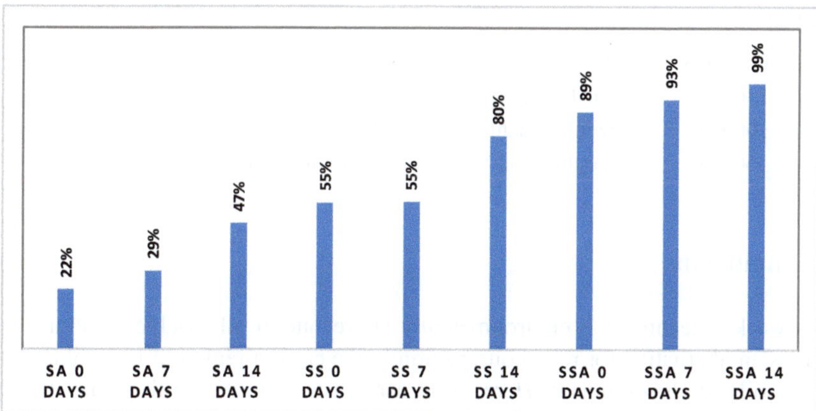

Fig. 6. Variation of SDI on slake – durability tests - Cycles 1 and 2.

Regarding the direct shear tests carried out, it was observed that ash addition does not develop significant increases in cohesion, but in friction resistance (See Fig. 7) providing values of friction angles from 27 to 35°, for earthworks design. Once again, there is evidence of an improvement in the soil resistance condition.

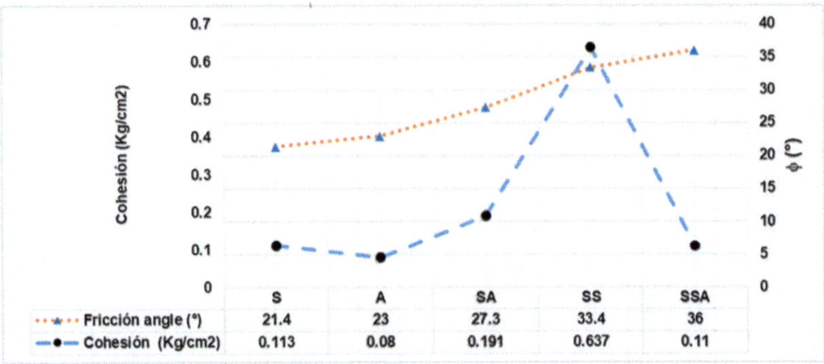

Fig. 7. Shear strength parameters C and ϕ for soil improvements

Complementary test for compacted ash showed CBR values close to 50%, but when mixed with the soil, the CBR goes down to 10%. Despite this, ashes require interaction with other agents to develop all the desirable properties in improved soils.

Finally, Table 1 presents a general estimate of costs for tertiary roads, considering just materials and transport cost for the region. The soil improvement with ash and silicate is the most economical alternative, compared with the conventional practices used.

Table 1. Estimated costs for tertiary roads (materials and transport - eastern region of Colombia)

Cost (US/km)	Materials
150000	Suitable, taken from remote areas
100000	Regular quality from local sources
60000	Marginal soil with Ash and Sodium Silicate improvement

4 Conclusions

In this work, a technically, environmentally and economically viable solution is presented as an alternative for road construction in the eastern region of Colombia, where lateritic soils are predominant. Here, solid ash waste from oil drilling cuts (ODC) is reused as a potential additive for clayey lateritic soil improvement to build road embankments or earthworks.

The complementary addition of sodium silicate to soil helps to create inter-particle bonding, increasing the compressive and shear strength.

Furthermore, the materials located along the corridors could be used for rural roads construction, reducing mining activity to obtain aggregates and the corresponding transport costs.

Based on the experimental campaign developed, it can be concluded that one of the main benefits achieved with the addition of ash to the lateritic soils of the region is durability, apparently due to the hydrophobic effect produced by a decrease in water migration by capillarity, probably related to reduction of surface tension forces.

These effects are reflected in a good performance before cycles of drying wetting and slake durability while increasing the resistance properties.

While, the addition of ash has favorable effects on soil durability, the addition of sodium silicate improves even more such properties and resistance. Therefore, it is recommended to apply ash and silicate in the proportions in weight identified in this work: soil 65%, ash 20%, 15% optimum compaction solution (3% sodium silicate and 12% water).

It is recommended to extend the experimental campaign to study the effect of applying some superficial waterproofing agent in preserving the improvements already achieved with the addition of ash and silicate to the soil.

Finally, building a road embankment test in the region is recommended to observe the behavior under actual traffic and environmental conditions.

Acknowledgements. The authors want to thank the support provided by the Department of Vichada to make this project possible. We also thank the students Brenda Sánchez and Jhonatan Suárez for helping in the experimental phase, and Oscar Caballero for sharing his results on soil improvement with sodium silicate.

References

Caballero, O.J.: Estabilización Química con Silicato de Sodio del Material de Préstamo de la Vía La Primavera – Bonanza – La Venturosa en el Departamento del Vichada (Chemical Stabilization with Sodium Silicate of soils from La Primavera - Bonanza - La Venturosa in the Department of Vichada. Master Thesis, National University of Colombia (2017)

Meegoda J.N., Muller R.T.: Petroleum contaminated soils highway construction. In: Symposium Proceedings: Recovery and in Effective Reuse of Discarded Materials and By-products for Construction of Highway Facilities, Denver, CO, pp. 483–495 (1993)

Tuncan, A., Koyuncu, H.: Use of petroleum contaminated drilling wastes as sub-base material for road construction. Waste Manag. Res. **18**, 489–505 (2000)

Pavement Friction Prediction for Single and Multiple Lane Rural Roads

Luciana Girardi Omar(✉) and Abd El Halim Omar Abd El Halim

Department of Civil and Environmental Engineering, Carleton University, Ottawa, ON, Canada
lucianaomar@cmail.carleton.ca, a_halim@carleton.ca

Abstract. Pavement friction is one of the most important operational requirements to ensure road safety. Adequate pavement friction is a fundamental requirement to perform safe driving, steering, and braking because it affects the interface between the tires of a vehicle and the pavement surface. Pavement friction is determined through a combination of two surface textures, microtexture and macrotexture. In Ontario, Canada, the provincial transportation agency is responsible for monitoring and maintaining adequate pavement friction; however, incorporating friction into the pavement management is still challenging because there is not a specific device that can measure both textures simultaneously at a high speed. This paper provides regression models that can be used to estimate skid resistance for rural roads based on field measurements of macrotexture, type of surface coarseness, aggregate properties, and operational condition of the roads. The models can be used by researchers and practitioners to predict skid resistance for asphalt pavements.

1 Introduction

Pavement texture is defined as "the deviations of the pavement surface from a true planar surface" (AASHTO 2008). Deviations are categorized into four levels, according to texture wavelength:

- Microtexture: wavelength < 0.5 mm;
- Macrotexture: wavelength > 0.5 mm to 50 mm;
- Megatexture: wavelength > 50 mm to 500 mm; and
- Roughness or unevenness: wavelength > 500 mm.

Pavement texture affects the interface between the tires of a vehicle and pavement surface. For instance, pavement friction is determined through a combination of two surface texture properties, microtexture and macrotexture (Kummer 1966). Megatexture and roughness are associated with ride-related qualities of pavement surface, such as pavement distresses, rolling resistance and level of noise, vibration, fuel consumption, and tire wear (AASHTO 2008; Noyce et al. 2007).

Adequate pavement friction is one of the most important operational requirements to ensure road safety. Pavement friction is a fundamental requirement to ensure safe driving, steering, and braking. However, incorporating friction into pavement

S. El-Badawy and R. Abd El-Hakim (Eds.): GeoMEast 2018, SUCI, pp. 24–39, 2019.
https://doi.org/10.1007/978-3-030-01908-2_3

management is still challenging for many transportation agencies, as there is not a specific device that can measure microtexture and macrotexture simultaneously.

In Ontario, skid resistance is a surrogate measurement for microtexture and is measured using a locked wheel trailer (LWT). Macrotexture, megatexure, and roughness are measured using a laser profilometer attached to high-speed automatic road analyzer (ARAN). The skid resistance measurements using LWT are relatively time-consuming and tend to be conducted less frequently than macrotexture measurements using ARAN because there is no system available that can measure microtexture at high speed due to the sensor's resolution and sampling limitations (Serigos et al. 2014). Therefore, there is a need to explore alternative approaches to correlating skid resistance and macrotexture measurements collected by automated road analysers, such as ARAN, that can be used by road agencies to detect friction deficiencies.

A problem that may arise in correlating microtexture measured with LWT and macrotexture measured with ARAN is that texture measurements do not always occur on the same lanes in multiple lane road segments. In multiple lane roads, the lane surveyed with LWT often does not correspond to the lane surveyed with ARAN. This difference may affect the correlation between them.

The objective of this study is to identify and investigate the main variables which govern the relationship between pavement skid resistance and macrotexture and to develop models for predicting skid resistance that correlate microtexture measured with LWT and macrotexture measured with ARAN. More specifically, the objectives of this study are: (i) to examine the variation of skid resistance and macrotexture due to the operational condition of the roads and type of mixes; and (ii) to develop a prediction models for estimating skid resistance for single and multiple lanes road segments that include macrotexture, type of surface coarseness, aggregate properties, and operational condition of the roads.

2 Background

Pavement friction is affected by two levels of texture, microtexture and macrotexture (Henry 2000). Microtexture is associated with the scale texture of the individual aggregate and depends on the physical nature of the aggregates. Poor levels of microtexture, resembling a smooth surface comparable to fine sandpaper, imply a low level of skid resistance at most speeds (Glennon and Hill 2004). Excellent levels of microtexture, resembling a grainy surface comparable to coarse sandpaper, allow a good level of skid resistance at low and high speeds except on flushing surfaces where there is no macrotexture (Glennon and Hill 2004). Microtexture provides sufficient friction on dry surfaces at all speeds and on wet surfaces at slower speeds (Hoerner and Smith 2002). Macrotexture is related to the height, width, angularity, and density of the macro projection of the aggregate above the surface (Hall et al. 2008). Macrotexture is usually assessed by the depth of the peaks of large aggregates. Macrotexture allows the drainage of water through channels formed by large aggregates or by grooves cut into the pavement surface. The level of pavement macrotexture is arguably correlated with the level of skid resistance, where a good level of macrotexture tends to give good levels of skid resistance at high speeds. Limited macrotexture associated with polished aggregate and combined with speed, tire pressure, and water depth increases the susceptibility of a vehicle to hydroplaning (Kummer 1966; Glennon and Hill 2004).

Macrotexture and microtexture are influenced by aggregates, binder, mix properties, and type of texturing applied over the surface. Pavement friction changes depending on its levels of microtexture and macrotexture (Leu and Henry 1978; Masad et al. 2007). Microtexture is mainly affected by the shape and mineralogy of coarse aggregates, while macrotexture is affected by other factors, such as maximum aggregate size, type of coarse and fine aggregates, mix gradation, compaction method, air content, and binder (Kandhal and Parker 1998). Higher levels of pavement surface texture are obtained when fine and coarse aggregates present irregular shapes and angular and cubical edges (Rado 2009). This improvement is attributable to the fractured faces of the coarse aggregates that improve the stability of the mixture and produce a deeper macrotexture (Shaffer et al. 2006).

The durability of frictional performance of pavement surface is affected by the polish and abrasion resistance of coarse and fine aggregates under traffic wear (Do et al. 2009). In general, friction increases in the initial few years due to wearing of the pavement surface by traffic and the exposure of rough aggregate surfaces, and then decreases at a constant rate over the remaining pavement life as aggregates become more polished (Kowalski et al. 2009; Mayora and Piña 2009). The presence of heavy load trucks and high traffic volume causes faster deterioration of frictional properties (AASHTO 2008; Ech et al. 2009).

2.1 Models for Predicting Pavement Friction

Several studies have estimated pavement friction using statistical models that correlate skid resistance and macrotexture with variables related to aggregate properties, mix design parameters, and road characteristics (e.g., traffic, loads, road design, and environmental conditions). For instance, Noyce et al. (2007) developed models of skid resistance based on macrotexture, speed, and mix design parameters of asphalt mixtures.

Ahammed and Tighe (2012) developed models for estimating skid resistance and macrotexture for asphalt pavement surfaces that correlated the ratio of the percentage of coarse and fine aggregates, speed, and aggregate quality. In another study, Rajaei et al. (2014) developed a model for predicting skid number and macrotexture using mix design parameters, such as bulk specific gravity, nominal aggregate size, binder content, and aggregate gradation that was defined by the two parameters of the Weibull distribution function (shape and scale parameters).

Other studies proposed models to predict skid resistance that correlated skid resistance and macrotexture measured using laser scanners. For instance, Serigos et al. (2014) developed a model to predict skid resistance accounting for the effects of macrotexture and microtexture spectral parameters calculated with the surface scanned by a laser texture scanner. Meegoda and Gao (2015) developed models to predict skid resistance of surfaces with different levels of macrotexture measured with a laser scanner attached to a high-speed road analyzer. Meegoda and Gao (2015) also proposed reduction factors for skid resistance to account for cumulative traffic volume according to different macrotexture levels.

In summary, a number of researchers have attempted to develop predicting models for skid resistance. It is observed from the abovementioned studies that models for estimating friction vary significantly depending on the type of equipment used for measuring pavement textures, the location where the textures were collected (field or laboratory), sample size, approaches, and the variables included in the models. It was also observed that limited studies attempted to develop predicting model for skid resistance using data collected at network level. Further, there is no study that addressed differences of measurements of skid resistance and macrotexture on different lanes. Thus, there is still a gap in the literature that needs investigation.

3 Study Sites and Data Collection

The data used in this research were obtained from the Ministry of Transportation of Ontario (MTO) for 1,363.64 km, distributed into 110 road segments of asphalt surface within 36 rural roads with speed limits of 80 km/h and 100 km/h. Of these 110 road segments, 53 were single lane road segments and 57 were multiple lane road segments. The field data, including skid resistance and macrotexture, were collected in the period from April to October of the years 2012 to 2015.

In addition to field data, information about the operational condition of the roads and the asphalt mix design was obtained for 50 road segments. Information on the operational condition of the roads included: years of service of the pavement (pavement age) after the most recent maintenance, rehabilitation, or construction; annual average daily traffic (AADT); annual average daily truck traffic (AADT truck); annual equivalent single-axle load traffic (AESAL); number of lanes; and lane distribution factor. Information about asphalt mix design parameters included: maximum nominal aggregate size (NMAS); percentage of voids in mineral aggregate (VMA); percentage of air void content (AV); percentage of asphalt content (AC); percentage of aggregates retained at sieve 4.75 mm (Co); the percentage of aggregates passing at sieve 4.75 mm (Fi); and percentage of aggregates passing at sieve 2.36 mm (FFi).

Information about the abrasion resistance and durability of coarse aggregates was given by the maximum percentage of Micro-Deval abrasion loss (MD). The MD percentage abrasion loss quantifies the resistance of aggregates to polishing. Table 1 displays a summary of the main parameters of the mix types.

The sample of 50 road segments included two section of hot mix HL type 4 (HL4), nine sections of Stone Mastic Asphalt (SMA), two sections of Stone Mastic Asphalt 12.5 (SMA12.5), 25 sections of Superpave 12.5 (SP12.5), 12 sections of Superpave 12.5 Friction Course 1 (SP 12.5 FC1), and seven sections of Superpave 12.5 Friction Course 2 (SP 12.5 FC2).

Skid resistance measurements occurred in the field on the left wheel path of the outside lane of travel using a locked wheel trailer (LWT) with a standard ribbed tire (ASTM E274/E274M 2015; ASTM E 501-97 2015). The LWT trailer was connected to a standard fleet truck and travelled along wetted surfaces at a constant speed of 64 km/h. The LWT trailer measured the force acting between the tire and the pavement when the braking system was activated to lock the tire. This force, also referred to as sliding force, was computed as the skid number (SN64R). The average of the skid

Table 1. Summary of mix parameters

Mix types	Mean of mix parameters						
	Coarse (%)	Fine (%)	VMA (%)	AC (%)	AV (%)	$NMAS_{max}$ (mm)	MD_{coarse} (%)
HL4	47.75	45.00	14.85	5.15	4.15	16.0	17.0
SMA	72.49	20.60	17.69	5.71	4.00	16.0	15.0
SMA 12.5	79.00	18.00	16.70	5.90	4.00	12.5	15.0
SP12.5	49.72	37.56	14.89	5.03	4.00	12.5	21.0
SP12.5 FC1	50.29	34.63	15.13	5.11	4.00	12.5	15.0
SP12.5 FC2	45.16	41.06	14.36	4.73	4.00	12.5	15.0

number was generated over 20 to 30 m of the road surface sample. On average, 13.8 skid resistance tests were performed per road segment, with a maximum of thirty-five tests in the longest segments and a minimum of four tests in the shortest segments. Curved road segments were not surveyed with the LWT. To determine the skid resistance level for each road section, the skid resistance measurements were computed as the average of the SN64R tests of each road section.

Macrotexture measurements occurred in the field using a high-speed road analyzer (ARAN), which is a multi-purpose and multi-sensor data collection vehicle. ARAN employs cameras, sensors, and an advanced laser scanner to survey pavement distress, macrotexture, smoothness, and other relevant attributes. The ARAN laser measurement system was based on regular contactless scanning of transverse and longitudinal sections of road. The lasers are also able to simultaneously measure both wheel paths of the lanes. The measurements were computed as average depths for intervals of 50 m length of the pavement surface. The longitudinal laser profiler measured the average depth of macrotexture, which was reported as Mean Profile Depth (MPD) (ASTM E1845-01 2015). To determine the macrotexture level for each road section, the macrotexture measurements were computed as the average of the MPD readings on the left wheel path of each road segment.

Matching pavement surface measurements of skid resistance and macrotexture presented a constraint: for 57 out of 110 road segments, the lane measured with the LWT did not correspond to the lane measured with ARAN. This occurred due to MTO operating procedures that state that measurements with the LWT regularly occur on the outside lane regardless of the number of lanes of the road section, while ARAN measurements may occur on the middle lane or inside lane, depending on the number of lanes in multiple lane road sections. Figure 1 shows a typical representation of measurements on a six-lane divided road section in which skid resistance is measured with the LWT on the left wheel path of the outside lane (lane 3), and macrotexture is measured with ARAN on the left and right wheel path of the middle lane (lane 2).

To estimate the AESAL on the lane surveyed with the LWT, the AESAL was adjusted according to the lane distribution factor (DF). The DF was based on the

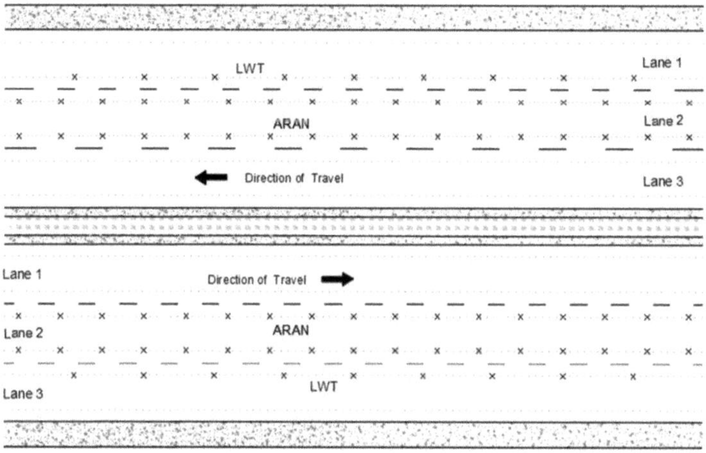

Fig. 1. Road segment layout

number of lanes in one direction, the percentage of trucks in the design lane, and the AADT truck (AADTt) (MTO 2012). The DFs were defined as follows:

- Two lanes in one direction: 0.8 (for AADTt < 15000) and 0.9 (for AADTt > 15000)
- Three lanes in one direction: 0.8 (for AADTt < 25000) and 0.7 (for AADTt > 25000)
- Four lanes in one direction: 0.7 (for AADTt < 40000) and 0.6 (for AADTt > 40000)

4 Preliminary Analysis

Statistical analyses were performed to examine the variance of SN64R and MPD due to the influence of age, traffic, and type of mix. The analysis was also categorized according to the number of lanes of road segments, i.e., single and multiple lanes. The confidence level used was 95%.

4.1 Variation of SN64R and MPD Due to the Pavement Age

The relationship between SN64R and MPD and pavement age were examined in terms of years of service after the most recent maintenance, rehabilitation, or construction. ANOVA results showed that there were statistically significantly differences between SN64R and MPD means after the third year for single and multiple lanes. Figure 2a and 2b show boxplots of the statistical distribution (minimum, median, quartiles, maximum, and outliers) of the SN64R and MPD of each category of pavement age. Figure 2a shows an increase of SN64R in the first three years, which was expected due to surface wearing that could be related to increased exposure of the fractured faces of aggregates that were initially covered by the asphalt binder. The following year, the fourth year, SN64R decreased, also as expected. The decreasing of SN64R could be attributed to the polishing effects of the aggregates due to traffic. From the period between the fifth year and the seventh year, contrary to what was expected, SN64R did not decrease at a constant rate or remain constant over the time; instead, SN64R

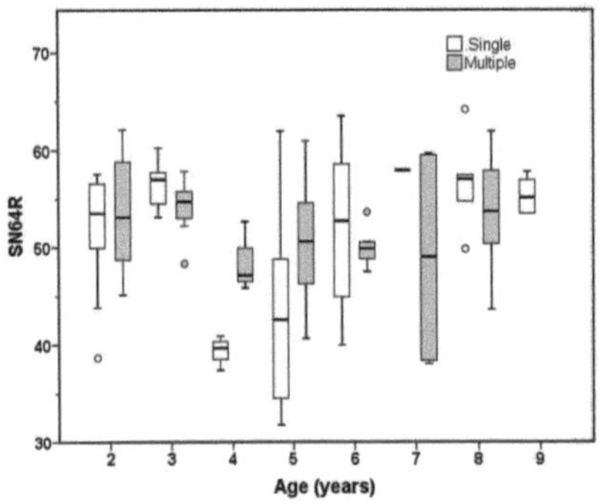

Fig. 2a. SN64R and pavement age

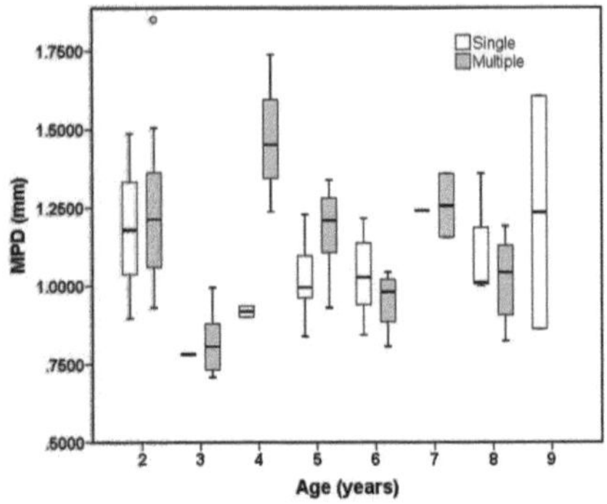

Fig. 2b. MPD and pavement age

gradually increased over these three years, and then exhibited a slight decrease in the eighth and ninth year.

As for SN64R, the data examined did not show a trend in changes of MPD over time. Figure 2b shows that MPD values alternated between increasing and decreasing over periods of two and three years for single and multiple lanes. Further, two-year-old pavements exhibited a similar MPD level as nine-year-old pavements for single lane road segments. In sum, the expected MPD increase over time due to the polishing effects of traffic was not observed in the data analyzed.

In conclusion, the effects of pavement age on SN64R and MPD for single and multiple lanes did not exhibit a consistent trend over the years. A possible explanation for the variation of SN64R and MPD values over time is that the cross-sectional data organization was not adequate to capture SN64R and MPD variations over time. It is possible that historical data of SN64R and MPD for each road segment would be more adequate to investigate SN64R and MPD performance over the years. Another possible explanation for the increase of friction in old pavements could be related to the increase of surface distresses, such as cracking and raveling that may contribute to an increase in roughness, as observed by Li et al. (2003).

4.2 Variation of SN64R and MPD Due to Traffic

The results indicated that SN64R and MPD were affected by traffic and loads. The results are displayed in Figs. 3a and 3b, which show the regression lines for SN64R and MPD and AESAL per lane and per type of road segments (i.e., single and multiple lanes). Figure 3a shows that SN64R decreases as the AESAL per lane increases for

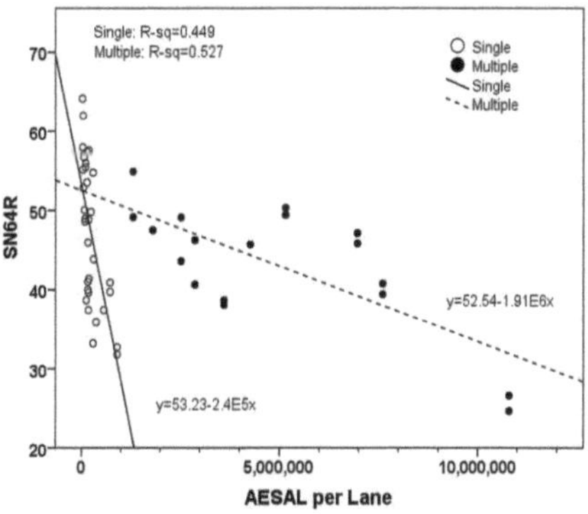

Fig. 3a. SN64R and AESAL per lane

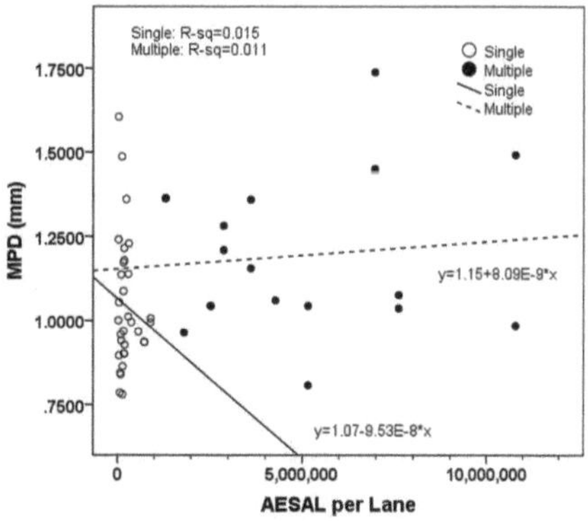

Fig. 3b. MPD and AESAL per lane

single and multiple lanes. The relationship between AESAL per lane and SN64R was greater for multiple lanes than for single lanes, with an R-square of 0.527 for multiple lanes and 0.449 for single lanes. Figure 3b shows that MPD decreases as the AESAL per lane increases for single lanes, and that MPD increases as the AESAL per lane increases for multiple lanes. However, the relationship between AESAL per lane and MPD was not significant for single and multiple lanes with an R-square of 0.015 for single lanes and 0.011 for multiple lanes.

4.3 Variation of SN64R and MPD Due to Type of Mix

Table 2 displays SN64R and MPD categorized per type of mix. ANOVA and Tukey post hoc test results indicated that SN64R means differed statistically significantly at a confidence level of 95% for the six types of mixes.

The SN64R means per type of mix and per type of road segment showed that SN64R means on single lane road segments were greater than on multiple lane road segments. For single lane road segments, the SP12.5 and HL4 exhibited greater SN64R

Table 2. SN64R and MPD per mix type and per road segment

Road segment	Mix type	N[a]	SN64R		MPD	
			Mean	SD	Mean	SD
Single	HL4	1	64.16	0.000	1.0000	0.0000
	SP12.5	20	49.09	8.944	1.0242	0.2024
	SP12.5 FC1	11	42.13	6.675	1.0788	0.1889
Means			47.17	9.171	1.0422	0.1933
Multiple	SP12.5 FC2	7	41.39	10.916	1.2252	0.3360
	SMA	9	42.69	4.093	1.1298	0.1305
	SMA12.5	2	52.04	4.043	1.3629	0.0000
Means			43.23	7.847	1.1928	0.2319

[a]N = number of road segments

means than SP 12.5 FC1. For multiple lane road segments, SMA12.5 exhibited greater SN64R means than SMA and SP12.5FC2, and SMA and SP12.5 FC2 exhibited comparable SN64R means. Results also indicated that for single lane road segments MPD means did not differ statistically significantly within the mixes. MPD means of multiple lane road segments were greater than MPD means of single lane road segments. For multiple lane road segments, SMA exhibited a lower MPD mean than SMA 12.5 and SP 12.5FC2, and SMA12.5 exhibited the highest MPD mean among the mixes.

5 Skid Resistance Modelling

For skid resistance modelling, this study used the ordinary least squares (OLS) statistical technique to investigate the relationships between a single dependent variable (SN64R) and several independent variables. In the OLS regression analysis each independent variable is weighted by the regression. The weights, represented by unstandardized coefficients (β, $\beta 1$, $\beta 2 \ldots$), denote the relative contribution of each independent variable to the overall prediction.

The set of potential predictors used in the modelling process included: macrotexture (MPD); pavement age (Age); traffic loads per lane (AESAL); nominal aggregate size (NMAS); percentage of voids in mineral aggregate (VMA); percentage of air voids (AV); percentage of asphalt content (AC); percentage of aggregates retained at sieve 4.75 mm (Co); percentage of aggregates passing at sieve 4.75 mm (Fi); and percentage of aggregates passing at sieve 2.36 mm (FFi).

Additionally, four other variables were created. The first variable addressed the cumulative effect of traffic over time, combining the effects of traffic by multiplying the variables AESAL by the age of the pavement (AgeAESAL). The second variable addressed the aggregate gradation distribution and was represented by the coefficient of the percentage of coarse and fine aggregates (Co/Fi). The third variable addressed the abrasion loss of coarse aggregates and was calculated by multiplying the percentage of coarse aggregates in the mix by the Micro-Deval abrasion loss for the respective percentage of coarse aggregate (CoMD).

The independent variable datasets were graphically analysed to meet three assumptions: normal distribution, homoscedasticity, and linearity. The frequencies of the data were plotted to verify whether data distribution follows a Gaussian distribution. The linearity was examined using standardized residual plots of residual versus the predicted dependent value. The independent variables that violated these assumptions were transformed using the logarithm function.

In total, 13 independent variables were tested and the variable selection procedure was maximized using a stepwise approach. This approach enabled the addition and removal of variables based on the contribution of each independent variable to the model. The modeling procedure started with a bivariate Pearson correlation matrix that displays combinations of dependent and independent variables. For each variable added to the model, several measures were analyzed to determine its contribution to the overall model fit regarding the increase in the R-square, decrease of standard error (SE), significance of the partial correlation (*p-value*), and the effects of collinearity. The effects of multicollinearity were assessed using the Variance Inflation Factor (VIF) with a threshold of five, which indicates a low level of collinearity. The variables with a higher VIF than the threshold were discarded.

5.1 Skid Resistance Models

An initial model for SN64R prediction was developed using the dataset from 50 road segments and the variables that showed great correlation with SN64R using the Pearson correlation coefficient. Among 13 variables tested, five variables were used to develop the model: MPD, AESAL, Co/Fi, CoMD, and Age. This initial model exhibited an R-square of 0.403 (*p-value* < 0.01). In this model, MPD, CoMD, and AESAL exhibited a negative correlation with SN64R, while Age and CoFi exhibited a positive correlation with SN64R. The correlation of MPD and AESAL with SN64R appeared to be logical based on the preliminary analysis. However, the positive correlation between SN64R and Age was illogical because SN64R is expected to decrease over time due to traffic polishing; therefore, the variable Age was excluded from the initial model. The variable CoFi also showed an illogical correlation with SN64R because in the analysis of the variation of SN64R and MPD per type of mix, the mix with a greater rate of coarse aggregates, SMA, exhibited similar SN64R to mixes with a lower rate of coarse aggregates, such as the superpaves. Further, the variable CoFi and CoMD exhibited a VIF greater than five, which indicated that they were multicorrelated. Thus, to eliminate the collinearity within independent variables, the variable Co/Fi was excluded from the initial model.

A second and a third model were developed using the three variables that were not excluded from the initial model (MPD, AESAL, and CoMD). In addition, the dataset was divided per type of road segment, i.e., single lane and multiple lanes. The model for single lanes exhibited a greater R-square than the primary model with an R-square of 0.664 (p-value < 0.01 and Standard Error = 5.61). The model for multiple lanes exhibited an R-square of 0.577 (p-value < 0.01 and Standard Error = 6.04).

$$\mathrm{SN64R_{RS}} = 140.92 - 16.09 \log \mathrm{AESAL} - 1.36\mathrm{MPD} - 3.05\mathrm{CoMD} \tag{1}$$

$$\mathrm{SN64R_{RM}} = 119.99 - 10.71 \log \mathrm{AESAL} - 3.55\mathrm{MPD} - 0.63\mathrm{CoMD} \tag{2}$$

where:

$SN64R_{RS}$ = Skid number measured with ribbed tire at 64 km/h for single lane,
$SN64R_{RM}$ = Skid number measured with ribbed tire at 64 km/h for multiple lanes,
AESAL = Annual ESAL per lane,
MPD = Macrotexture in mm,
CoMD = Percentage of coarse aggregates and percentage of Micro-Deval aggregate loss of coarse aggregates.

The models indicated that the R-squares and MPD unstandardized coefficients differed in terms of value for single and multiple lanes. The model for single lanes exhibited a greater R-square than the model for multiple lanes. The model for single lanes also exhibited an AESAL unstandardized coefficient greater than the model for multiple lanes. This difference in the AESAL unstandardized coefficients may be related to the influence of estimating AESAL per lane using the DF for multiple lane road segments.

The model for single lanes also exhibited an MPD unstandardized coefficient lower than the model for multiple lanes. The differences in the MPD unstandardized coefficients may be related to differences in MPD due to the measurements being taken in different lanes.

The models also indicated that the CoMD unstandardized coefficients differed in terms of values within the models. The CoMD unstandardized coefficient in the model for multiple lanes is lower than the coefficient in the model for single lanes. This difference may be related to differences in the percentage of coarse aggregates and percentage of abrasion loss of the coarse aggregates. For instance, in the SN64R model for single lanes, the reduction of SN64R in mixes with similar percentage of coarse aggregates, such as SP 12.5 and SP 12.5 FC1, is greater in the mix with greater MD, i.e., SP 12.5 with MD of 21%. This result seems to be logical because the lower MD indicates a better quality of aggregates.

5.2 A Model with a Categorical Variable

A third model, Model 3, was developed to predict SN64R for single and multiple road segments using a single regression equation that incorporated a categorical variable to represent single and multiple lane road segments. In Model 3, an additional dummy variable (SM) with 0 (zero) for single lane and 1 (one) for multiple lanes was included. Model 3 exhibited an R-square of 0.591 (*p-value* < 0.01 and Standard Error = 5.90). Model 3 is shown in Eq. 3.

$$\mathrm{SN64R_{RSM}} = 121.59 - 12.93\log\mathrm{AESAL} - 1.68\mathrm{MPD} - 1.89\mathrm{CoMD} + 13.63\mathrm{SM} \quad (3)$$

where:

$\mathrm{SN64R_{RSM}}$ = Skid number measured with ribbed tire at 64 km/h,
AESAL = AESAL per lane,
MPD = Macrotexture in mm,
CoMD = Percentage of coarse aggregates and percentage of Micro-Deval aggregate loss of coarse aggregates,
SM = Dummy variable with 0 for single lane and 1 for multiple lanes.

Figure 4 shows three regression lines of Model 3, the overall regression line, and two separate regression lines, where one line shows when the model implies SM is equal to 0 (single lane) and another line shows when SN64R is equal to 1 (multiple lanes). Figure 4 also shows that the R-square for single lanes (R-square = 0.661) was greater than the R-square for multiple lanes (R-square = 0.565). Further, Model 3 exhibited R-squares for single and multiple lanes comparable to the R-squares of Model 1 and Model 2. Thus, Model 1 and Model 2 can be replaced by Model 3.

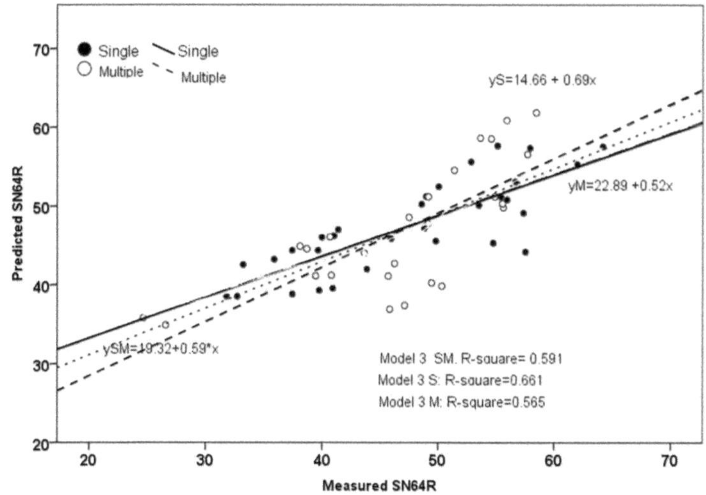

Fig. 4. Regression lines of model 3

5.3 Model Evaluation

The residuals of the regression model, Model 3 was verified by plotting the predicted SN64R values and the standardized residuals. Figure 5 shows a nonlinear pattern and the points are randomly distributed with an even spread of residuals at all predicted values, which indicates that the model meets the assumption of linearity.

The normality of the errors of residuals was examined using observed cumulative and the expected cumulative proportion. Figure 6 shows the expected and observed cumulative probabilities fairly match, which indicates that the regression meets the assumption of normality. The analysis of multicollinearity indicated that all variables in the model exhibited VIF lower than five.

The accuracy of the regression model, Model 3, was evaluated using the bootstrapping statistical technique of resampling the same data sample study using the percentile method with a confidence interval of 95%. The bootstrap sample with 95% percentile confidence interval was calculated using the range of the bootstrap values corresponding to the 2.5th and 97.5th percentiles. The bootstrap values were generated with 1000 bootstrap replications using the statistical software SPSS 23 (IBM 2015). The model obtained using the bootstrap resampling exhibited an R-square of 0.574 (*p-value* < 0.01 and Standard Error of 6.21) which was analogous to the original model.

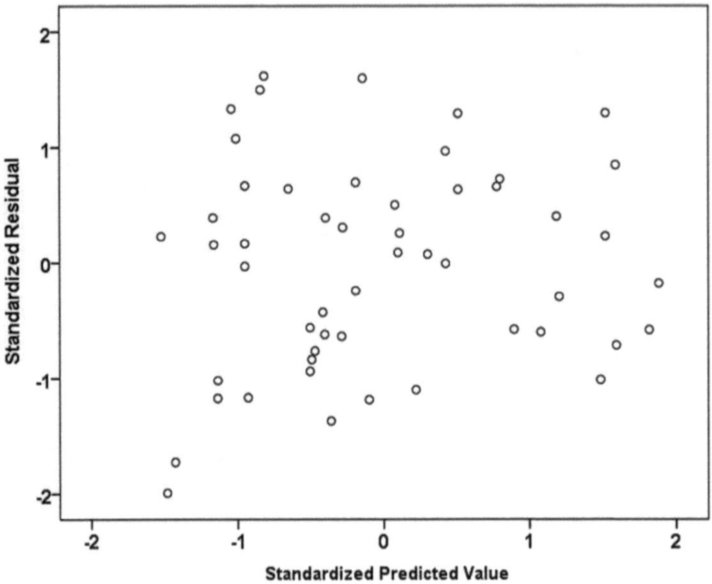

Fig. 5. Standardize residuals of model 3

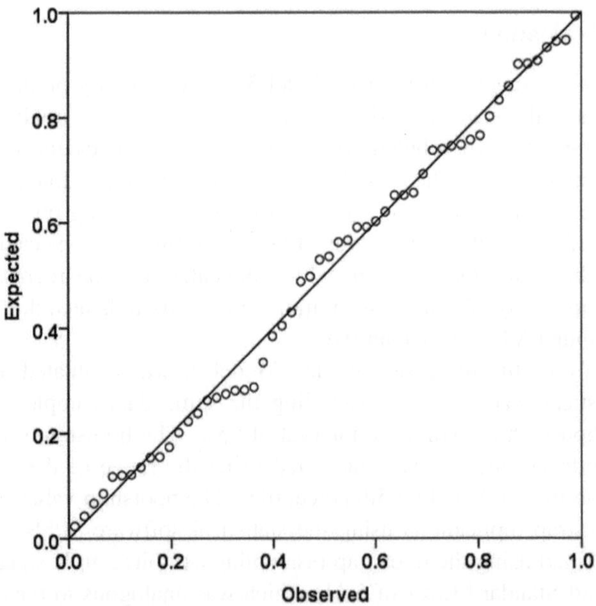

Fig. 6. Normal probability of residuals of model 3

6 Conclusions

This study examined the variation of skid resistance (SN64R) and macrotexture (MPD) due to the effects of pavement ages, type of mix, and traffic. Three prediction models for estimating SN64R were developed for single and multiple lane road segments. The main findings of this research were as follows:

- The data examined did not show a trend of SN64R decreasing over time. SN64R showed an increase in the first three years, which was expected due to wearing of the surface; after three years, SN64R alternated between increasing and decreasing until the end of the period of the ninth year. As for SN64R, the data examined did not show a trend in changes of MPD over time. MPD was expected to increase over time as a result of the polishing effects due to the traffic; however, this trend was not confirmed. Similar to SN64R, MPD alternated between increasing and decreasing over periods of 2 and 3 years. One possible explanation for these results is that the cross-sectional approach to data organization may not have been suitable for this type of analysis.
- SN64R differed noticeably across different types of mixes. The superpave mixes and HL4 mix exhibited SN64R means higher than SMA. MPD did not differ in the same magnitude within the type of mixes; however, SMA 12.5 exhibited the highest levels of MPD and HL4 and SP 12.5 exhibited the lowest levels of MPD.
- SN64R and MPD were affected by traffic and loads. SN64R decreased as the AESAL increased and MPD increased as the AESAL increased.

The modelling results show that the model for SN64R prediction for single lane road segments exhibited a greater R-square than the model for multiple lane road segments. This finding supports the assumption that measurements of SN64R and MPD on the same lane increase model accuracy. In the regression models, MPD, AESAL, and CoMD exhibited a negative correlation with SN64R, which indicates that an increase of macrotexture and traffic and loads, and an increase of the percentage of coarse aggregates with high percentage of abrasion loss negatively influence skid resistance.

The skid resistance prediction models developed in this study could be a useful tool to estimate skid resistance from data collected at network level for the Ontario roads. The prediction models could be incorporated into pavement management to trigger treatments to maintain adequate skid resistance levels and improve road safety.

The data available for this study was limited to a range of years and highways. It is possible that underlying relationships between variables were not detected due to this limitation. Additional research should investigate skid resistance variance over time. It is recommended to use historical data obtained from the same sites for a sequence of years, instead of cross-sectional data obtained from many sites within few years. It is also recommended to include pavement condition in the analysis to evaluate the influence of pavement distresses, such as ravelling and cracks, on skid resistance. Additionally, more data should be collected to validate the SN64R prediction models for single and multiple lanes.

Acknowledgments. This research was supported by the Ministry of Transportation of Ontario, Highway Infrastructure Innovation Funding Program and the Natural Science and Engineering Research Council of Canada.

References

AASHTO: Guide for Pavement Friction. American Association of State Highway and Transportation Officials, Washington (2008)

Ahammed, M.A., Tighe, S.L.: Asphalt pavements surface texture and skid resistance-exploring the reality. Can. J. Civ. Eng. **39**(1), 1–9 (2012)

ASTM E1845-15: Standard Practice for Calculating Pavement Macrotexture Mean Profile Depth. ASTM International, West Conshohocken (2015)

ASTM E274, E274 M–15: Standard Test Method for Skid Resistance of Paved Surfaces Using a Full-Scale Tire. ASTM International, West Conshohocken (2015)

ASTM E501-08: Standard Specification for Standard Rib Tire for Pavement Skid-Resistance Tests. ASTM International, West Conshohocken (2015)

Do, M.-T., et al.: Evolution of road-surface skid-resistance and texture due to polishing. Wear **266**(5–6), 574–577 (2009)

Ech, M., et al.: An original evaluation of the wearing course macrotexture evolution using abbot curve. Road Mater. Pavement Design **10**, 471–494 (2009)

Glennon, J.C., Hill, P.F.: Roadway Safety and Tort Liability. Lawyers & Judges Publishing Company, Tucson (2004)

Hall, J.W., et al.: American Association of State Highway and Transportation Officials, Washington, DC (2008)

Henry, J.J.: Evaluation of Pavement Friction Characteristics. National Academy Press, Washington, D.C. (2000)

Hoerner, T.E., Smith, K.D.: High Performance Concrete Pavement: Pavement Texturing and Tire-Pavement Noise. Federal Highway Administration, Washington, D.C. (2002)

IBM: IBM SPSS Statistics for Windows, Version 23.0. IBM Corp, Armonk (2015)

Kandhal, P., Parker, F.L.: Aggregate Tests Related to Asphalt Concrete Performance in Pavements. Transportation Research Board, Washington, DC (1998)

Kowalski, K., et al.: Long-term monitoring of noise and frictional properties of three pavements. Transp. Res. Rec. J. Transp. Res. Board **2127**, 12–19 (2009)

Kummer, H.W.: Unified theory of rubber ad tire friction. Eng. Res. Bull. **B-94**, 135 (1966)

Leu, M.C., Henry, J.J.: Prediction of Skid Resistance as Function of Speed from Pavement Texture Measurements. Transportation Research Record: Journal of the Transportation Research Board: Journal of Transportation Research Board, No. 666, pp. 7–13 Transportation Research Board of the National Academies, Washington (1978)

Li, S., et al.: Upgrading the INDOT Pavement Friction Testing Program. Indiana Department of Transportation, Indianapolis (2003)

Masad, E., et al.: Relationship of Aggregate Texture to Asphalt Pavement Skid Resistance Using Image Analysis of Aggregate Shape. Transportation Research Board, Washington, DC (2007)

Mayora, J.M.P., Piña, R.J.: An assessment of the skid resistance effect on traffic safety under wet-pavement conditions. Accid. Anal. Prev. **41**(4), 881–886 (2009)

Meegoda, J.N., Gao, S.: Evaluation of pavement skid resistance using high speed texture measurement. J. Traffic Transp. Eng. **2**(6), 382–390 (2015). (English Edition)

MTO: Ontario's Default Parameters for AASHTOWare Pavement ME Design - Interim Report. Ontario Ministry of Transportation, Downsview (2012)

Noyce, D.A., et al.: Incorporating Road Safety into Pavement Management: Maximizing Surface Friction for Road Safety Improvements. Midwest Regional University Transportation Center, Madison (2007)

Rado, Z.: Evaluating Performance of Limestone Prone to Polishing. The Pennsylvania Department of Transportation, Harrisburg (2009)

Rajaei, M., et al.: Establishment of relationship between pavement surface friction and mixture design properties. Transp. Res. Rec. J. Transp. Res. Board **2457**, 114–120 (2014)

Serigos, P., et al.: Incorporating Surface Microtexture in the Prediction of Skid Resistance of Flexible Pavements. Transp. Res. Rec. J. Transp. Res. Board **2457**, 105–113 (2014)

Shaffer, S.J., et al.: Assessment of Friction-Based Pavement Methods and Regulations. National Transportation Research Center, Incorporated, Knoxville, TN, p. 133 (2006)

RETRACTED CHAPTER: Improvement of Rheological Properties of Asphalt Binder by Adding Composite Montmorillonite Nanoclay

Abdulhaq Hadi Abedali Al-haddad(✉) and Rana Amir Yousif

Al-Mustansiriyah University, Baghdad, Iraq
abdulhaq1969@uomustansiriyah.edu.iq

Abstract. Various concentration of composite nanoclay was studied to determine the physical and rheological properties of three types of asphalt binders. One type of composite Nanoclay in different concentrations (0%, 3%, 5% and 7%) by weight of bitumen were used and blended with three asphalt binder types (80/100-60/70 and PG-76|) penetration grade. The physical properties of unaged base and nanoclay modified asphalt were characterized using viscosity, softening point, and penetration tests. Dynamic Shear Rheometer were used to determine the rheological properties of unmodified and nanoclay modified asphalt in accordance with AASHTO T315 in order to evaluate the effectiveness of composite nanoclay type and concentration on the physical and rheological properties of asphalt binder. The results indicated remarkable increment in softening point; viscosity and decrement in asphalt binder's penetration for both types and contents of organic nanoclay and asphalt types as compared with unmodified bitumen. The results of the experiments indicated that the addition of nano particles was helpful in increasing the complex modulus values and in improving rutting resistance of the RTFO binder. The phase angle of the binders generally decreased with an increase in nano content and RTFO aging procedure. Also, the results indicated that modified binders show better rheological properties compared to standard bitumen for all asphalt type. The addition of Nanoclay to base asphalt binder has led to noticeable improvements in aging resistance this may be due to the homogeneous dispersion of nano particles consisting of layer silicate in the asphalt medium. The main objective of this study is to study the effect of nanoclay on the physical and rheological properties for three different type of binder.

Keywords: Bitumen · Nanoclay · Modified asphalt · Physical and rheological properties

1 Introduction

The focus on the efficiently of pavement structure is the most important issue that attract the interesting of the road engineers as the paved road reflect one of the progress features of the countries. It is well known that the pavement using the conventional asphalt binder safer from many distress during its construction and service life. The fatigue, rutting, thermal cracking and moisture induced damage are the most familiar

The original version of this chapter was retracted: The retraction note to this chapter can be found at https://doi.org/10.1007/978-3-030-01908-2_16

S. El-Badawy and R. Abd El-Hakim (Eds.): GeoMEast 2018, SUCI, pp. 40–53, 2019.
https://doi.org/10.1007/978-3-030-01908-2_4

distress that threatened the pavement structure (Lottman 1978). In order to improve the asphalt properties, many of additives were adopted. Polymer, rubber, fiber and filler are the most known modifiers that used to enhance and reinforce the asphalt properties (Sargand and Kim 2003; Little and Epps 2001; Liu et al. 2009; Yildrim 2007). But, the thinking towards new materials which can collect new promising properties such as the lower price and the environmental friendly in addition to improving the binder and mixture properties, still occupied the researcher's thinking. Nanotechnology using nano material now a day, is the most interesting technology that adopted by mostly the industries and academics and showed unexpected reactive by the community. Nanomaterials have at least one dimension measuring less than 100 nm (nm) at least. Due to their small size, usually nanomaterials have the higher reactivity ability and special surface properties, which can be used for industry products. Due to the special properties of the above-mentioned, nanomaterials, they are good candidates for implementation into asphalt pavements. In recent studies, it was found that nanoclay could increase the shear complex modulus and reduce the strain failure rate of base asphalt where nanoclay material was added to modify the base asphalt binder. Furthermore, the moisture susceptibility of asphalt mixture could weaken by adding the nanoclay in the base asphalt binder (You et al. 2011).

Two types of nanoclay, nanofill and cloisite were tested by Ghile, were the various properties of asphalt binder and mixture studied including indirect tensile strength, dynamic creep, and fatigue resistance for mixture (Ghile 2006). Two studies introduced by Jianying Yu, where the effect of two types of nanoclays, inorganic montmorillonite and organic montmorillonite that used to modify bitumen investigated, one of the study explored the effect of types of montmorillonite on properties of binder and the other showed the effect of Montmorillonite (MMT) and OMMT on oxidation temperature and on binder aging properties (Jianying et al. 2007, 2009). Nanofil-15 and Cloiste-15A were also adopted by Jahromi, properties like the stiffness, and aging resistance where explored (Jahromi and Khodaii 2009). Shaopeng Wu, has chosen Organic Montmorillonite (OMMT) to prepare the nanoclay modified bitumen, his study carried out to evaluate the fatigue resistance property of the modified bitumen (Wu et al. 2010). One Organic Montmorillonite was selected by Gang Liu with different contents, where the thermal properties and morphology studied before and after aging (Liu et al. 2009). All the above research results indicated that the addition of nanoclay improves different properties of bitumen with respect to stiffness, rutting, aging and fatigue to some degree. In this study one 80/100 penetration grade asphalt binder was chosen and two types of nanoclay with three different contents were used to investigate their effect on physical and theological properties of asphalt binder. Saeed et al. (2009) studied the effect of nanoclay on the internal structural of the blends while.

2 Materials and Methods

2.1 Materials

2.1.1 Asphalt Binder

Three asphalt penetration grades were used in this study (80-100, 60-70 and PG-76). The bitumen used was provided by Kajang Rocks Innopave Premix Company (KRIP) in Malaysia.

2.1.2 Composite Nanoclay

One type of organic montmorillonite nanoclay, were obtained from USA. Code 682624. The study was done in two stages; in the first stage the effect of nanoclay and content on the physical properties of unaged asphalt binder was studied using three asphalt binder 80-100, 60-70 and PG-76 penetration grade; the second stage was done to evaluate the rheological properties of unaged nanoclay modified and unmodified asphalt binder.

2.1.3 Preparation of Composite Nanoclay Modified Asphalt

To prepare the nanoclay modified asphalt for physical and rheological testing, asphalt binder was heated in an oven until melted and poured in a container. The blend was prepared by adding the predetermined amount of nanoclay material gradually around 1 g/min to the hot binder at established mixing temperatures of (150 ± 5 °C) at speed of 500 rpm for 80-100 asphalt grade (160 ± 5 °C) at speed of 500 rpm for 60-70 asphalt grade and (165 ± 5 °C) at speed of 500 rpm for PG-76 asphalt grade; the blend was mechanically mixed on an electric hot-plate set using IKA Labortechnik, RW 20 DZM.n mechanical mixer. After all the nanoclay materials was added to the asphalt binder, the material was stirred for 60 min at speed of 2000 rpm while maintaining a temperature of 150 °C to ensure that the nanoclay particles were homogenously and well dispersed inside the medium of bitumen binder.

2.1.4 Physical Properties Test of Composite Nanoclay Modified Asphalt

The physical properties tests of the base and composite nanoclay modified asphalt that carried out including penetration according to ASTM D 5, Softening Point according to ASTM D 36, and Viscosity using Brookfield viscometer at two different temperatures of 135 °C and 165 °C according to ASTM D 4402.

2.1.5 Rheological Properties of Composite Nanoclay Modified Asphalt

Dynamic Shear Rheoemeter (DSR) (HAAKE, RheoStress RS1, Phoenix) has been adopted for measuring the rheological properties of unaged base and composite nanoclay modified binder. The DSR is used to measure the rheological prosperities of asphalt binders at high and intermediate temperatures. The DSR measures the complex shear modulus, G* value and phase angle δ of the binder at the desired temperature and loading frequency. Complex shear modulus G* can be considered as the total resistance of the binder to deformation when repeatedly sheared. The phase angle represents the immediate elastic and the delayed viscous responses of the binder obtained from the lag between the measured shear stresses and the induced strains in a strain-controlled test.

For elastic materials, the phase angle value is zero, whereas for purely viscous materials, the phase angle is 90 °C. Thus, the phase angle is important in describing the visco-elastic properties of a material such as asphalt (The Asphalt Institute 2007; Richard Kim 2009).

2.2 Rutting Performance

The Superpave rutting parameter (G*/sin δ), was identified as a term to be used for high temperature performance grading, usually larger than 46 °C, of paving asphalts in rating the binders for their rutting resistance. Basically, (G*/sin δ), was recommended as the Superpave specification parameter to provide a measure of the rutting resistance of asphalts. The higher the G* value, the stiffer and thus the more resistant to rutting the asphalt binder is. The lower the δ value, the more elastic the binder, knowing that an increase in elasticity makes the asphalt binder more resistant to permanent deformation. Strategic Highway Research Program (SHRP) researchers consider that, with each traffic load cycle, work is being done to deform the Hot Mix Asphalt (HMA) pavement surface. Part of that work recovered due to the elastic stored energy if it has enough time, while the remaining work is dissipated through permanent deformation that contribute to the rutting in pavement layers (Bahia and Anderson 1995; Bahia 2001). In order to minimize permanent deformation (rutting), the amount of work dissipated during each loading cycle must be minimized (NCHRP Report 465 2002). It has been found out that the rutting susceptibility should decrease with the increase of the (G*/sin δ) values, thus the Superpave parameter is intended to control rutting by controlling the total energy dissipated per cycle which can be calculated from the strain–stress curve, using Eqs. (1) and (2).

$$\varepsilon = \frac{\sigma}{G^*} \tag{1}$$

$$W_c = \pi\sigma^2\left(\frac{1}{G^*/sin\delta}\right) \tag{2}$$

Where

W_c = work dissipated per load cycle
σ = stress applied during load cycle in kPa
ε = strain during load cycle
G* = complex shear modulus
δ = phase angle in degree.

2.3 Dynamic Shear Rheometer Testing Procedures

The DSR rutting factor, G*/sin δ, was used to measure the intermediate and high temperature stiffness and rutting resistance of asphalt binders. The SHRP Superpave PG binder specifies minimum values for G*/sin δ of 1000 Pa for original asphalt binder, 2200 Pa after Rolling Thin Film Oven (RTFO) aging and 5000 kPa after Pressure Aging Vessel (PAV) procedure (AASHTO TP 5-93). After placing the

OMMT modified specimen into the DSR device, Bohlin software was used to perform the test oscillation. Using 25 mm diameter spindle and gap of 1 mm, the unaged asphalt sample is compressed between two parallel plates, one of which is fixed and the other is oscillates. The temperature steps test was conducted between 40 and 82 °C with 6° increment, at fixed frequency of 10 rad/s in controlled stress mode of 120 Pa. in accordance to Superpave specification.

3 Results and Discussion

This section presents analysis and discussion of the results, which are obtained from the physical tests and the Dynamic Shear Rheometer (DSR) for unaged base and OMMTs modified binder at three different concentrations, one loading frequency and eight different temperatures. The tests results were used to evaluate the improvements of physical and rheological properties; complex shear modulus, phase angle, and rutting parameter of asphalt binder with respect to concentration proportions (by weight of asphalt) of nanoclay.

3.1 Physical Properties Test Results

The results of the laboratory tests of physical properties of asphalt binder and nanoclay types are shown in Tables 1 and 2.

Table 1. Physical properties of base asphalt 80/100, 60/70 and PG-76 penetration grade.

Physical properties	Result			Specification
	80-100	60-70	PG-76	
Penetration@ 25 °C, 0.1 mm	84	63.5	33.5	ASTM D 5
Softening Point, °C	46.5	49	73.5	ASTM D 36
Viscosity@ 135 °C, Pa s	0.3	0.410	1.480	ASTM D 4402
Viscosity@ 165 °C, Pa s	0.098	0.125	0.478	ASTM D 4402
Ductility @ 25 °C, cm	>100	>100	>100	ASTM D 113
Specific gravity, g/cm^3	1.03	1.04	1.03	ASTM D 70

Table 2. Description of nanomaterial.

Physical properties	Nanoclay
Organic montmorillonite treatment	Quaternary ammonium compounds, bis (hydrogenated tallow alkyl) dimethyl, chlorides.
CEC	35–45 wt% dimethyl dialkyl (C14-C18) AMINE MATRIX
Color	White
Appearance	Fine
Moisture	3%

The results of the laboratory tests of rheological properties of asphalt binder and OMMTs modified binder containing different content of Nano clay are presented in Figs. 1, 2, 3, 4, 5, 6, 7, 8, 9, 10, 11, 12 and 13. The consistency of the binder was measured using the empirical physical tests: penetration and softening. The variation of penetration and softening point value between unmodified and various concentrations of OMMTs modified asphalt binder are shown in Figs. 1 and 2. The viscosity using Rotational Brookfield Viscometer was carried out on unmodified and OMMT modified asphalt binder and the test results are shown in Figs. 3 and 4.

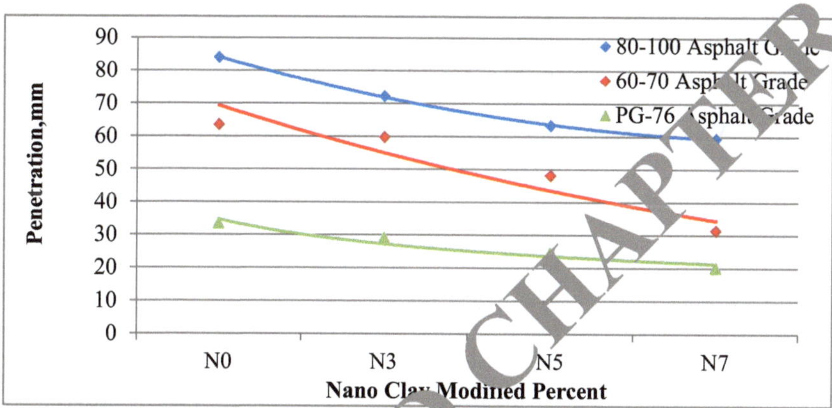

Fig. 1. Penetration value for three types of nano-clay modified asphalts.

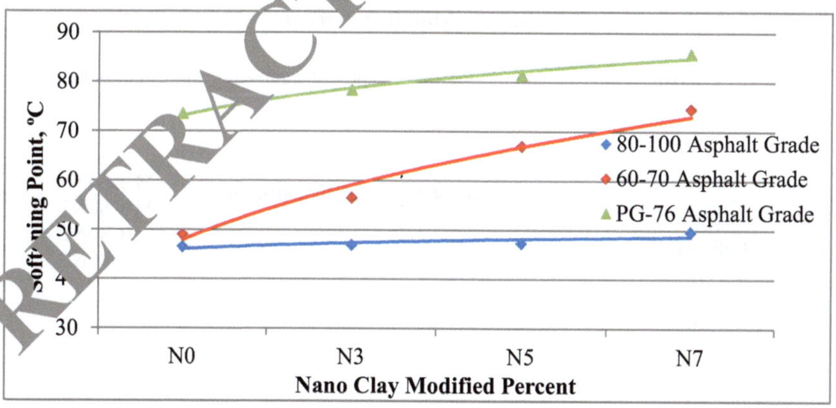

Fig. 2. Softening point value for three types of nano-clay modified asphalts.

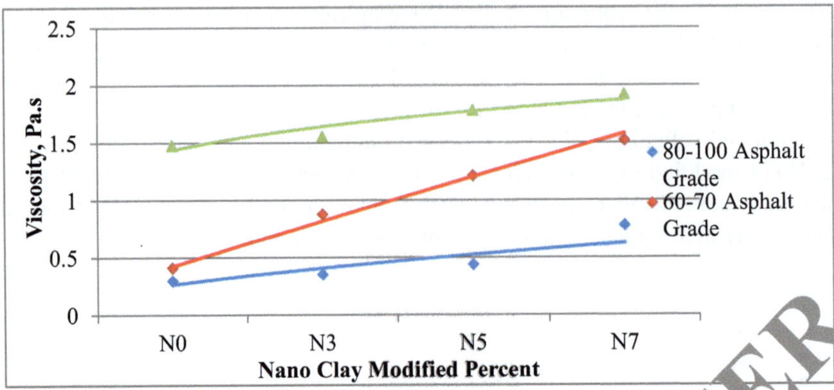

Fig. 3. Viscosity value at 135 °C for three types of nano-clay modified asphalts

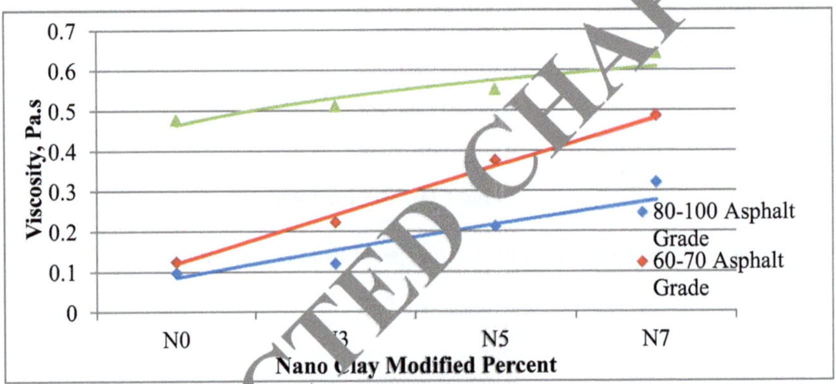

Fig. 4. Viscosity value at 165 °C for three types of nano-clay modified asphalts

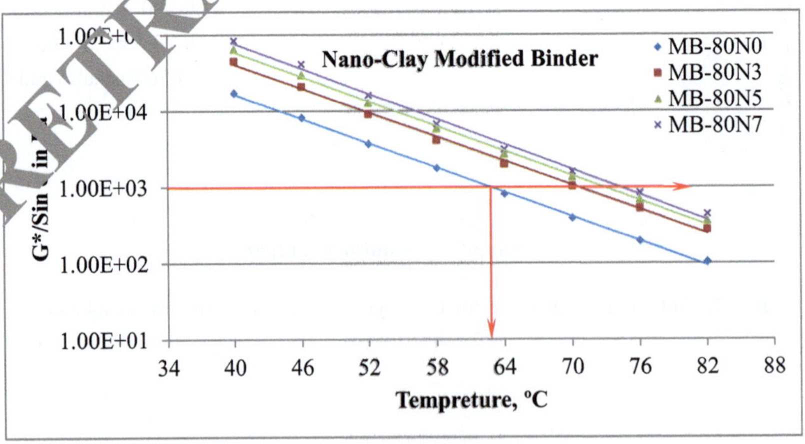

Fig. 5. DSR test result for 80-100 asphalt binder with nano-clay modified binder

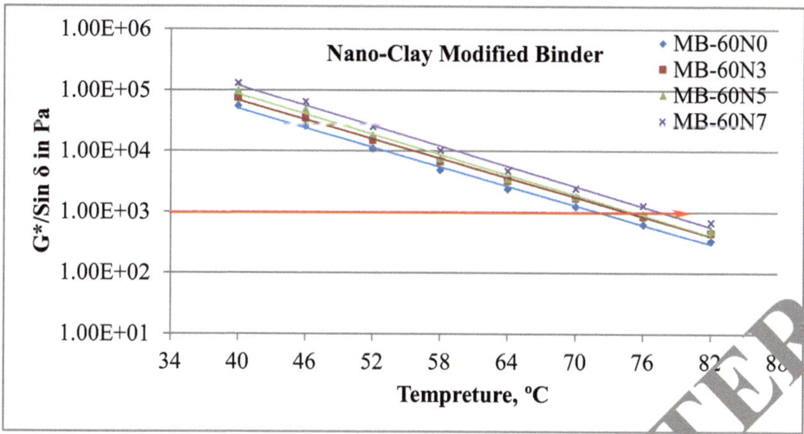

Fig. 6. DSR test result for 60-70 asphalt binder with nano-clay modified binder

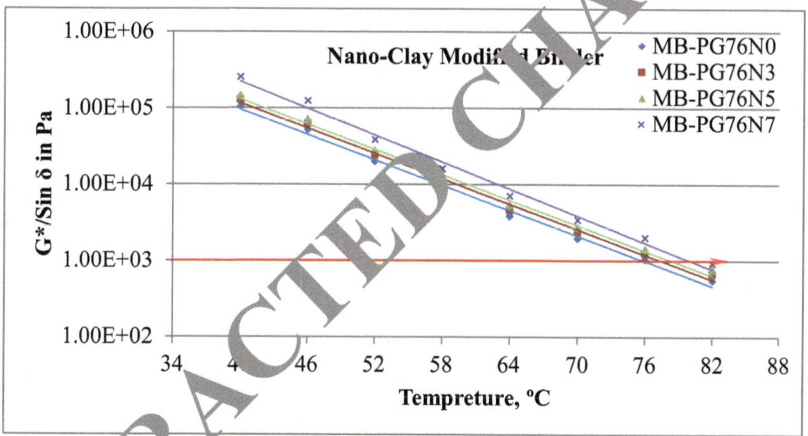

Fig. 7. DSR test result for PG-76 asphalt binder with nano-clay modified binder

It can be noticed from Figs. 1, 2, 3 and 4 that there is a remarkable improvement in penetration and softening point of the Nanoclay modified bitumen binder compared to the base asphalt binder. The addition of Nanoclay to base asphalt have led to decrease in penetration, increase in softening point and viscosity at both testing temperatures, this improvement is due to the increased amount of nanoclay content in the binder for the three type of binder. The main reason of this improvement after adding the various OMMT concentrations might be due to the better dispersion of the OMMT layers in the binder which increased the bonding strength through restrict the flow of bitumen made it stiffer and led to improve its physical properties.

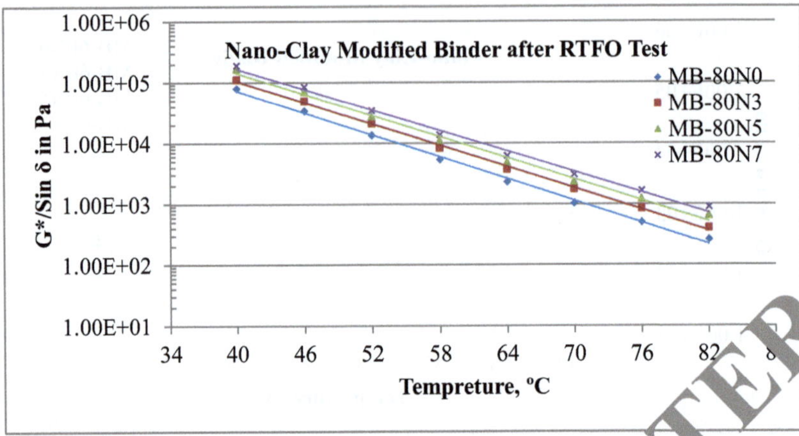

Fig. 8. DSR test result for 80-100 asphalt binder with nano-clay modified binder after RTFO test

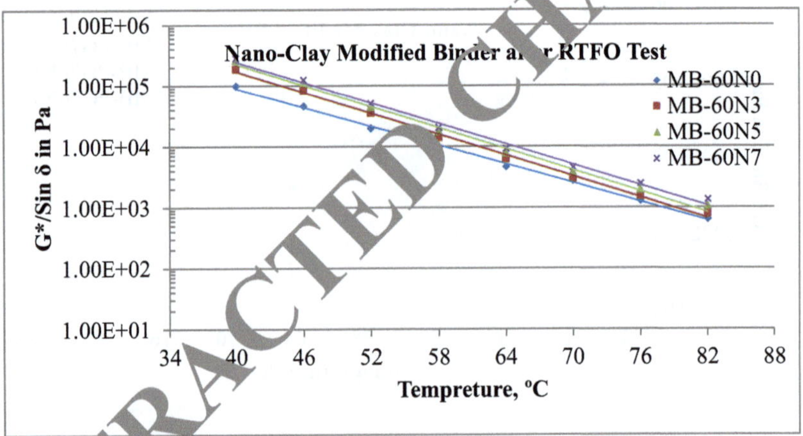

Fig. 9. DSR test result for 60-70 asphalt binder with nano-clay modified binder after RTFO test

3.1.1 Temperature Steps Tests of OMMT/Modified Binders

The temperature steps tests were performed under stress control and frequency of 10 rad/s (1.592 Hz) and employed a range of temperature from 40 to 82 °C. The overall temperature steps tests were run with a 25 mm diameter and 1 mm testing gap geometry complex modulus and phase angle were evaluated along with elastic and viscous modulus in accordance with various temperatures. Figure 7 displays the influence of temperature variation on the rutting parameter (G*/sinδ) values of various samples from base and Nanoclay modified binder (in unaged and aged states) at 40–82 °C. An increase in temperature yields decrease in the modulus values for base and Nanoclay modified binders, regardless of aging state or the presence of nanoparticles.

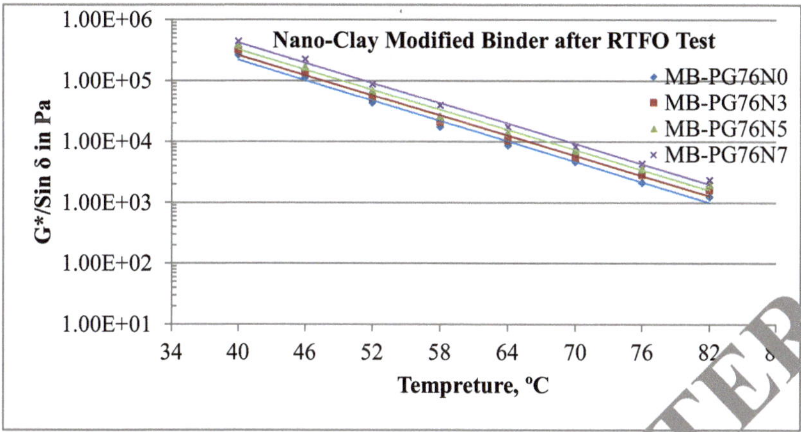

Fig. 10. DSR test result for PG-76 asphalt binder with nano-clay modified binder after RTFO test

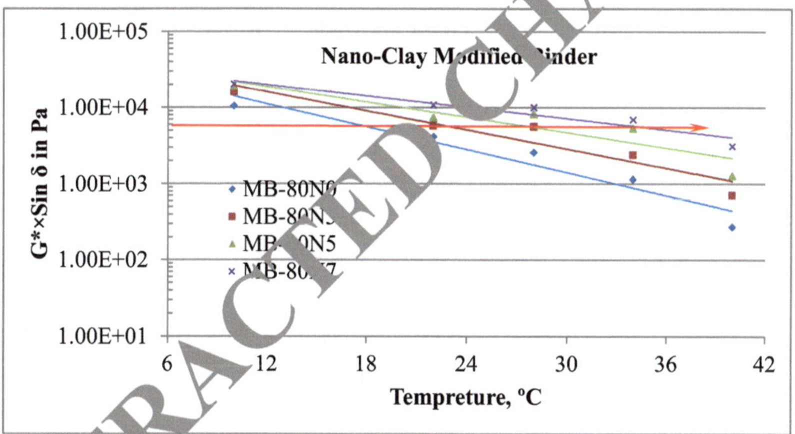

Fig. 11. Time sweep test for 80-100 asphalt binder with nano-clay modified.

Moreover, the RTFO samples generally have higher rutting parameter values than unaged samples.

3.1.2 Failure Temperature

Figures 5, 6, 7, 8, 9 and 10 illustrates that addition of nanoparticles results in an increase in failure temperature for all modified binder specimens, regardless of the aging state. However, this increase is generally different for various RTFO binder specimens. As presented in Table 3, it can be noted that for RTFO aged nanocomposite nanoclay content increased by four grade i.e. the high temperature grade was one grade higher than the base asphalt binder for all types of asphalt. A general trend was observed that the rutting parameters increased as the amount and concentration of

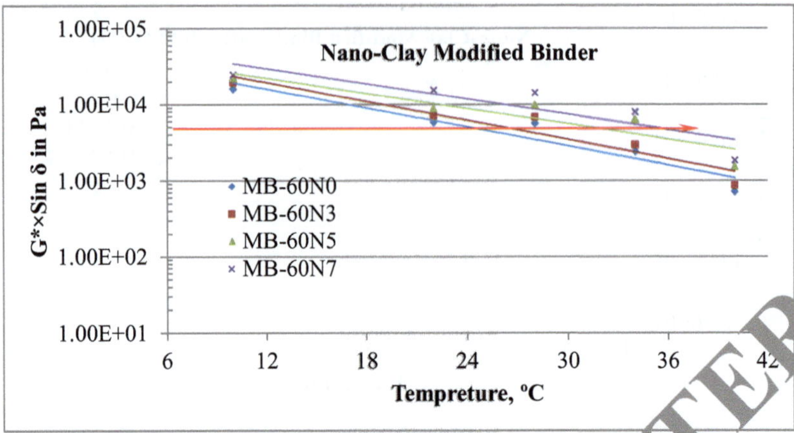

Fig. 12. Time sweep test for 60-70 asphalt binder with nano-clay modified.

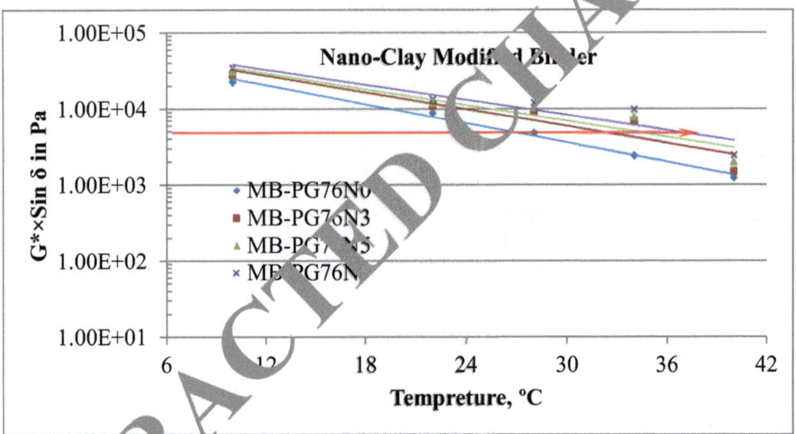

Fig. 13. Time sweep test for PG76 asphalt binder with nano-clay modified.

nanoclay increases as illustrated in Figs. 5, 6, 7, 8, 9 and 10 which reflects the increment of the binder resistance to deformation under repeated load. The results of DSR tests also showed that PG-76 gave higher G*/sin δ values than 80-100 and 60-60 at the three selected nanoclay contents which indicated better performance of PG-76 when permanent deformation is concern, this can be contributed to the type of surface modifier that chosen to treated the montmorillonite nanoclay to convert it to organic.

3.2 Time Sweep Test Results

Time sweep test protocol was similar to asphalt fatigue testing method for the asphalt binder using DSR. This test was ideal to observe how asphalt binder changes over time. This test was carried out on RTFO aged samples with a diameter of 8 mm and

Table 3. Failure temperature and PG-grading of nano-clay modified asphalt binder

Binder type	Failure temperature (°C)			Temperature high and low PG – grading
	Un-aged	RTFO aged short term aging	PAV aged long term aging	
MB-80N0	62.5	64.5	18.2	PG-58-19
MB-80N3	71.0	68.5	22.5	PG-64-23
MB-80N5	72.7	71.3	27.5	PG-70-28
MB-80N7	75.0	74.5	34.0	PG-70-34
MB-60N0	71.3	71.5	24.5	PG-70-25
MB-60N3	74.0	74.4	26.5	PG-70-28
MB-60N5	75.0	75.2	31.7	PG-70-33
MB-60N7	77.2	77.5	35.5	PG-76-3[illegible]
MB-PG76N0	76.5	76.8	27.3	PG-[illegible]-28
MB-PG76N3	77.5	77.8	32.2	[illegible]-76-[illegible]
MB-PG76N5	79.5	80.5	34.4	PG-76-36
MB-PG76N7	81.5	81.8	37.0	PG-76-38

thickness of 2 mm. the requirement temperature was 20 °C with the frequency of 10 Hz. The results of time sweep test are shown in Figs. 11, 12 and 13.

The increment of failure temperature is very important to improving the high performance of asphalt binder. It will help to reduce the high-temperature deformation and rutting damage of asphalt pavement. Table 3 below shows the Failure Temperature and PG-Grading for Nanoclay Modified binder at three different proportions laboratory prepared blown asphalt binder at five different temperatures and steam aged binder at five different times.

4 Conclusions

Based upon the analysis and results of this study, the basic conclusions can be drawn:

1. The physical properties of composite nanoclay modified asphalt binder showed tremendous enhancement as compared to base asphalt binder.
2. From the DSR test results, it can be concluded that the Complex Shear Modulus (G*) increased and the Phase Angle (δ) decreased compared to the base asphalt which led to remarkable improvement in Rutting Factor (G*/sin δ), in reducing the energy dissipated per load cycle and enhance the recovery ability as the amount of composite nanoclay increased in asphalt binder which result in a stiffer binder as result of dispersion of nanoclay particles in bitumen which made it less deformed and consequently reduced rutting susceptibility.
3. The increase in the complex shear modulus value as a result of the addition of nanoclay could be used to improve the Superpave high temperature Performance Grade (PG) value of asphalt binders.

References

ASTM D 5: Standard Test Method for Penetration of Bituminous Material, Volume 04.03. Annual Book of ASTM Standards, American Society for Testing and Material (ASTM), West Conshohocken (1986)

ASTM D 36: Standard Test Method for Softening Point of Bituminous Material, Volume 04.03. Annual Book of ASTM Standards, American Society for Testing and Material (ASTM), West Conshohocken (1986)

ASTM D 70: Standard Test Method for Density of Semi-Solid Bituminous Materials (Pycnometer Method). American Society for Testing and Materials (ASTM), West Conshohocken (1997)

ASTM D 113: Standard Test Method for Ductility of Bituminous Materials. United States: American Society for Testing and Materials (1999)

(ASTM), USA. ASTM D 4402: Standard Test Method for Viscosity Determination of Unfilled Asphalts Using the Brookfield Thermo sell Apparatus, Volume 04.03. Annual Book of ASTM Standards, American Society for Testing and Material (ASTM), West Conshohocken (1987)

AASHTO, TP 5-93: Standard Test Method for Determination of Rheological Properties of Asphalt Binder Using a Dynamic Shear Rheometer. American Association of State Highway and Transportation Officials (AASHTO), Washington, D.C. (1995)

Bahia, H.U., Anderson, D.A.: The SHRP binder rheological parameters: Why are they required and how do they compare to conventional properties. TRB, Preprint paper No. 950793 (1995)

Bahia, H.U., Zhai, H., Zeng, M., Hu, Y., Turner, P.: Development of binder specification parameters based on characterization of damage behavior. J. Assoc. Asphalt Pav. Technol. **70**, 442 (2001)

Liu, G., Wu, S., van de Ven, M., Molenaar, A., Besamusca, J.: Modification of bitumen with organic montmorillonite nanoclay. AES-ATEMA 2009. In: Third International Conference on Advances and Trends in Engineering Materials and Their Applications/Montreal, Canada (2009)

Ghile, D.B.: Effects of nanoclay modification on rheology of bitumen and on performance of asphalt mixtures. Master Thesis, The Delft University of Technology, Netherlands (2006)

Jahromi, S.G., Khodaii, A.: Effects of nanoclay on rheological properties of bitumen binder. Constr. Build. Mater. **23**, 2894–2904 (2009)

Yu, J., Feng, P.-C., Zhang, H.-L., Shao-Peng, W.: Effect of organomontmorillonite on aging properties of asphalt. Constr. Build. Mater. **23**, 2636–2640 (2009)

Yu, J., Zeng, X., Wu, S., Wang, L., Liu, G.: Preparation and properties of montmorillonite modified asphalts. Mater. Sci. Eng. A **447**, 233–238 (2007)

Liu, S., Cao, W., Fang, J., Shang, S.: Variance analysis and performance evaluation of different crumb rubber modified (CRM) asphalt. Constr. Build. Mater. **23**(7), 2701–2708 (2009)

NCHRP Report 465: Simple Performance Test for Superpave Mix Design. National Center for Asphalt Technology, Washington, D.C. (2002)

Wu, S., Wang, J., Jiesheng, L.: Preparation and fatigue property of nanoclay modified asphalt binder. In: International Conference on Mechanic Automation and Control Engineering (MACE) (2010)

SHRP-A-410: Superior Performing Asphalt Pavements (Superpave): The Product of the SHRP Asphalt Research Program. Strategic Highway Research Program. National Research Council, Washington, DC (1994)

The Asphalt Institute: The Asphalt Handbook. Manuel Series No: 4 (MS-4), 7th Edn. USA (2007)

Richard Kim, Y.: Modeling of Asphalt Concrete. Book of McGraw-Hill Publishers, 1st edn. American Society of Civil Engineers, Reston (2009)

Yildrim, Y.: Polymer modified asphalt binder. Constr. Build. Mater. **21**, 66–72 (2007)

You, Z., Mills-Beale, J., Foley, J.M., Roy, S., Odegard, G.M., Dai, Q., Goh, S.W.: Nanoclay modified asphalt materials: preparation and characterization. Constr. Build. Mater. **25**(2), 1072–1078 (2011)

Lottman, R.P.: Predicting Moisture Induced Damage to Asphalt Concrete. NCHRP Report 192, Transportation Research Board, Washington, D.C. (1978)

Sargand, S.M., Kim, S.-S.: Performance evaluation of polymer modified superpave mixes using laboratory tests and accelerated pavement load facility. In: 82nd Transportation Research Board Annual Meeting, Washington, D.C. (2003)

Little, D.N., Epps, J.A.: The benefits of hydrated lime in hot mix asphalt. Report for National Lime Association (2001)

Improving Mechanical Properties of Hot Mix Asphalt Using Crumb Rubber in Libya

Khlifa El Atrash(✉) and Gabriel J. Assaf

École de Technologie Supérieure, Université de Québec, Montreal, Canada
k_atrch@yahoo.com

Abstract. This paper deals with the possible reuse of rubber waste such as car tires and plastic bottles in the design of asphalt mixtures in Libya. This work is a continuation to the laboratory research that began in the last year. In the earlier study, the volumetric properties of the traditional asphalt mixture using penetration grade binder (B60/70) were compared with the Superpave mixture using performance grade binder (PG70-10). However, this study is to evaluate the effect of the crumb rubber (CR) on the performance of the asphalt mixture in Libya. Four asphalt concrete mixtures were designed using asphalt-binder B60/70, aggregate and CR and then tested by both Marshall and Superpave mix design procedures. These results were compared with the results obtained from the previous traditional Marshall mix design test using asphalt binder B60/70 and Superpave mix design test using PG70-10. This study clearly showed that asphalt mixtures with CR performed slightly better than the traditional mixtures with binder B60/70. In addition, the results obtained from the Superpave mix design test with asphalt binder PG70-10 have shown a better performance indicator than those obtained with Marshall modifier mixture with CR. These results provide a guide line to improving the production and mechanical properties of the hot mix asphalt (HMA) in Libya by adopting new mix design methods such as Superpave or adding additives such as CR to the traditional mix design method with binder B60/70.

Keywords: Asphalt-modifiers · Performance · Additives
Marshall mix design method · Asphalt binder

1 Introduction

Improving the mechanical properties and long-term characteristics of hot mix asphalt (HMA) should be considered as an intention to achieve a truly sustainable method of infrastructure development. However, this issue becomes a challenge if you are using a conventional HMA. In fact, the performance of a conventional HMA typically represents a poor long-term performance and functional problems associated with the daily traffic intensity and extreme temperature changes, which in turn implies higher maintenance costs. One of the methods that should be used in Libya to improve the behavior of HMA is the use of an asphalt binder that is highly sensitive to the temperature changes such as modified bitumen, performance grade binder or use of additives to the asphalt mixture. On the other hand, the preparation of hot mix asphalt

S. El-Badawy and R. Abd El-Hakim (Eds.): GeoMEast 2018, SUCI, pp. 54–63, 2019.
https://doi.org/10.1007/978-3-030-01908-2_5

by adding CR can have very interesting advantages in terms of economy, production and sustainability (Zhu et al. 2014). These advantages will be discussed in more detail in the next sections of this paper. Thus, laboratory tests including Marshall and Superpave mix design were performed to assess the effect of the inclusion of CR on the mechanical and volumetric properties of the asphalt mixtures. The results showed that the CR can be used to improve the characteristics and long-term properties of HMA. Asphalt rubber is a noise-reducing material for paving and improves the flexibility of asphalt pavement which consists of conventional asphalt mixed with rubber chips from recycled tires (Park et al. 1996). Asphalt mixture properties such as density and air voids are largely dependent on the type of the asphalt binder, compaction method, selection of aggregate and method of mix design. These properties, in turn, have an influence over the pavement performance indicators such as rutting, fatigue, potholes and alligator cracks (El Khatib 2016). The properties of the asphalt mixture depend very much on its temperature. At high temperatures, asphalt mixtures become viscous and behave like a plastic body under load traffic that exceeds its viscosity at a certain temperature. This behavior at high temperature can be a contributing factor to one of the most common distresses of asphalt pavement, which is rutting. In extremely cold climates, asphalt mixtures become very rigid and behave like an elastic body in which induced elastic deformation is completely restored. It is characterized by several failures represented by the low temperature cracking, fatigue cracking, and shrinkage cracks (Haixu et al. 2012). Therefore, this research aims to improve the mechanical properties of the HMA mixture to obtain an economical mixture that would meet the requirements for the characteristics of the asphalt roads in Libya. The Marshall Mix Design Method used the asphalt binder B (60/70) based on the empirical Penetration Grade System. It is not capable of viscosity measurements and hence, it is difficult to establish the correct HMA. Therefore, this paper presents results of laboratory tests for four specimens of asphalt mixtures. Two asphalt concrete mixtures which are designed using asphalt-binder B60/70, aggregate and CR and then tested by Marshall mix design procedure, the results compared with results obtained from the previous traditional Marshall mix design test using asphalt binder B60/70. Another two asphalt concrete mixtures were designed using asphalt binder B60/70, aggregate and CR and then tested by Superpave mix design procedure, the results compared with the previous results obtained from the Superpave mix design using asphalt binder PG70-10.

2 Objectives

Asphalt pavement in Libya is significant and constantly increasing, and the use of additives such as solid industrial wastes (SIW), polymer (BSR), performance grade binder (PG) and CR as components in new mixtures is strongly supported by companies who produce the asphalt. Currently, the specification of the asphalt binder in Libya is based on the Penetration Grade, which is an empirical measure of consistency and is almost always used as a traditional indicator of the susceptibility to deformation (rutting) and fatigue; it is not related to the pavement performance. To overcome these weaknesses, the bitumen industry has been developing new modifiers and additives to improve HMA behaviour. Therefore, this research aims to improve the mechanical properties of hot mix asphalt using additives in Libya to obtain an economical mixture that would meet the requirements of the pavement in hot weather.

3 Background and Literature Review

There was fairly extensive experience of using CR in developed countries, some with a wet process, others with dry. More than 3,000 km of streets in USA were paved with asphalt rubber mixtures in the early 1970s. The use of chip-seals was discontinued in the 1990s in favor of single-layer coatings based on 1-inch asphalt rubber. A superposition of a hot mixture is used to pave about 600 roads in USA. It is stated that both chip seals and hot mix linings can be used to slow the reflection of alligator cracks and shrinkage cracks with a width of less than 6.3 mm (1/4 inch). It is also concluded that, the asphalt-rubber on the hot mix allows for a more improved riding surface and a significant reduction in traffic noise (Charania et al. 1992). The Washington State Department of transportation (WSDOT) has used three types of paving from a wet process since 1977. Detergent technology products include SAM, SAMI courses and open asphalt-rubber friction courses (Kaya 1992). WSDOT reported that the performance of asphalt rubber SAM and SAMI was not sufficient enough to validate their construction's extra costs (Salini 2000). All five units with an open friction course demonstrate good performance, except for one deck overlap, which displays some discomfort in the wheel path area (Heitzman 1992). Three demonstration projects with rubber modified asphalt have been evaluated in Ontario, Canada in terms of road surface performance. The success of projects with asphalt-rubber (wet process) was encouraging, as the strength of these asphalt mixtures seems to have been improved by using a crumb rubber modifier. Actions to modify asphalt concrete material began in the 1840s. Unfortunately, the goal was difficult to achieve, the formulas of asphalt-rubber were not successful, and the result was a modified asphalt that was more expensive to produce and difficult to maintain than the usual asphalt (Roy et al. 2013). The Marshall mix design method was initially developed by Bruce G. Marshall from the Mississippi Highway Department in 1939, and later amended by the US Waterways Experiment Station of the US Army to include deformation measurements. This is based on two criteria, a minimum stability and a minimum range of flow values. However, it was based on a single piece of equipment to design and control the asphalt paving mixtures which are referred to as the Marshall Test (Almadwi and Assaf 2017). By 1968, the Arizona Department of Transportation in the United States began a variety of research and development projects involving asphalt rubber. By 1975, crumb rubber (CR) was successfully incorporated into HMA (Amorim et al. 2015). In the mid-1980s, the Europeans began developing new polymers and additives for use in modifying the bituminous binder (Brule 1996).

4 Methodology

Asphalt concrete mixtures were designed during laboratory work using asphalt-binder B60/70, aggregate and CR and then tested by Marshall and Superpave mix design procedures. These mixtures were used to evaluate the performance of the bituminous mixtures in the region. Also, the specimens were used for a comparison of several asphalt mixture using Marshall and Superpave mixtures with the traditional mix design methods in Libya. Evidently, the test mixtures contained several asphalt bitumen contents both above and below the optimum content of asphalt.

4.1 Mix Design Experiments

The purpose of this research is to determine the proper modified proportions of aggregates and asphalt to obtain an economical asphalt mixture that would meet the requirements of pavement in hot weather. Over the years, several design and development methods have been taking place and have been implemented by various agencies.

4.2 Materials Used

Aggregate
In this research, the mix of coarse and fine aggregate with sizes between 0.08 to 14 mm was used in this mix design according to the requirements. Tables 1, 2, and Fig. 1, show the characterises and gradation of the aggregate.

Table 1. Sieve analysis result and aggregate gradation for laboratory mix design

Sieve size mm	% Lower limit	% Upper limit	% Passing
20	95	100	98.7
14	67	90	85.6
10	52	75	68.1
5	34	55	44.87
2.5	24	45	27.6
1.25	16	39	17.7
0.63	9	31	12.7
0.31	6	23	9.68
0.16	4	15	7.55
0.08	3	8	6.33

Bituminous Binder
In this research, the asphalt binder for asphaltic concrete mix was the binder grade PG70-10 or B60/70.

Table 2. Specific gravity for the compound of aggregates and asphalt binders.

Agg. size and asphalt binder	% of Agg. in mixture	Bulk specific gravity	Compound mix specific gravity	% Water absorption
20–14 mm	15	2.739	2.716	0.630
14–10 mm	15	2.736		0.670
10–5 mm	26	2.733		0.700
Crushed 0–5 mm	36	2.698		0.005
Sand 0–5 mm	3	2.595		0.007
Filler mm	5	2.700		0.006
Bitumen (B60/70)	4.50	1.020	1.020	-
Bitumen PG 70–10	4.25	1.290	1.029	-

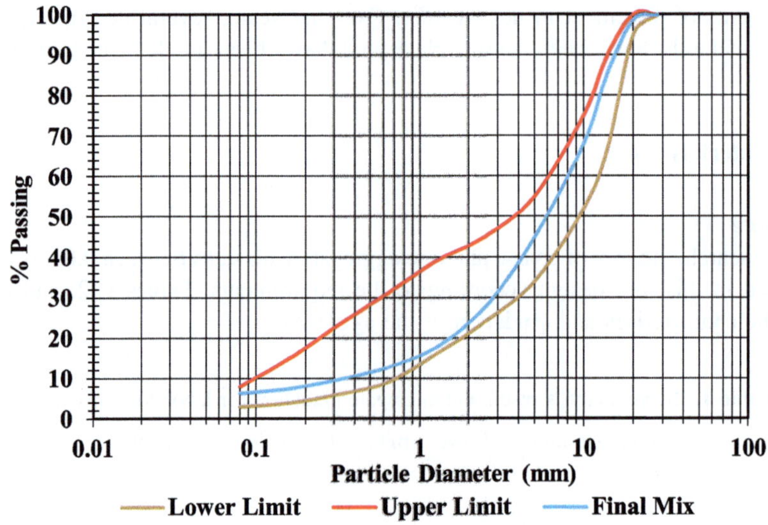

Fig. 1. Aggregate grading curve

Mineral Filler

The mineral filler used for this study was the dust of limestone. It should be dry and essentially free of agglomerations.

Crumb Rubber

The rubber from scrap can be included in asphalt mixes using two methods, a wet process and a dry process. In the wet process, crumb rubber performs as an asphalt cement modifier, while in the dry process, granular rubber or crumb is used as part of the aggregate skeleton. In both cases, crumb rubber is sometimes referred to as a rubber modifier (RM) and in most cases the output is called the modified mixture, which is usually used in a content between 7% to 22% of bitumen, (Chesner et al. 1998). The manufacturers are giving the information and recommendation on the use of the modifier. Based on the manufacturer's instructions in this experiment, the CR was added to hot aggregates with a dose 0.5% by weight of the dry aggregates prior the addition of bitumen.

5 Laboratory Experiments Results

Mix Design Test

In this experiment, specimens were produced by using aggregate, mineral filler, asphalt binder B60/70 and CR. The specimens were then tested by both Marshall and Superpave mix design procedure, and the results obtained are presented in Tables 3 and 4 illustrated all the laboratory experiment results. Figures 2 and 3 show the behavior of the modifier-asphalt mixtures compared to traditional mixtures. However, in this process before mixing, the aggregates were heated to 110 °C to ensure that there was no water content. Then aggregates were heated until they reached the mixing temperature (179 °C) for asphalt modified mixtures (Zoorob and Suparma 2000). This temperature was chosen in accordance to the CR data sheet. The percentage of additives was chosen by trial and error method; 10%, 15%, 20% were tried and finally 18% was determined to be the best ratio, which gave the best homogeneous mixture. The modified additive compound CR was then poured into a mixer and mixed with the aggregate.

Table 3. Result of Marshall specimens with two different mixtures

Mixture properties	Unmodified mixes with B (60/70)				Modified mixes with B (60/70)			
	SI	SII	SIII	Ave.	SI	SII	SIII	Ave.
OBC%	5.0	5.2	5.5	**5.2**	6.5	7.0	7.5	**7.0**
VMA%	14.5	15.5	16.6	**15.6**	15.9	18.7	19.0	**17.8**
VFA%	68.4	62.9	63.9	**65.1**	60.6	68.1	63.5	**64.0**
Va%	5.2	5.8	6.5	**5.8**	7.4	5.8	7.6	**6.9**
Flow (mm)	4.1	3.6	3.9	**3.8**	4.2	3.5	3.2	**3.6**
Stability (KN)	9.5	9.4	8.6	**9.2**	22.2	18.3	16.7	**19.1**

Modified Mixture Test Results Using Superpave Mix Design Procedures

In this experiment, specimens were prepared using asphalt binder B60/70, CR with 18%, filler and aggregate, and the other specimens were tested in the pervious laboratory work with the same aggregate gradation, filler and asphalt binder PG70-10. The obtained results at gyrations numbers 10, 80, and 200 are shown in Table 4 and illustrated in Fig. 3.

Modified Mixture Test Results Using Marshall Mix Design Procedures

In this section, all specimen experiments were conducted by using Marshall Mix design methods including modified additives CR. Marshall Mix design method was used for modified mixtures to determine the optimum bitumen content (OBC). Samples with standard sizes were compacted with a Marshall hammer with 75 blows on each side and at (168 °C). Mechanical and volumetric properties of all samples were obtained

Table 4. Result of Superpave specimens with two different mixtures

Numbers of gyrations	Superpave mixes with PG 70-10			Modified mixes with B (60/70)		
	VMA %	VFA%	Va%	VMA %	VFA%	Va%
10	23.21	52.63	12.21	24.4	51.53	11.75
80	20.45	62.45	6.54	18.35	60.54	5.95
200	21.65	70.02	4.12	18.64	70.53	3.25

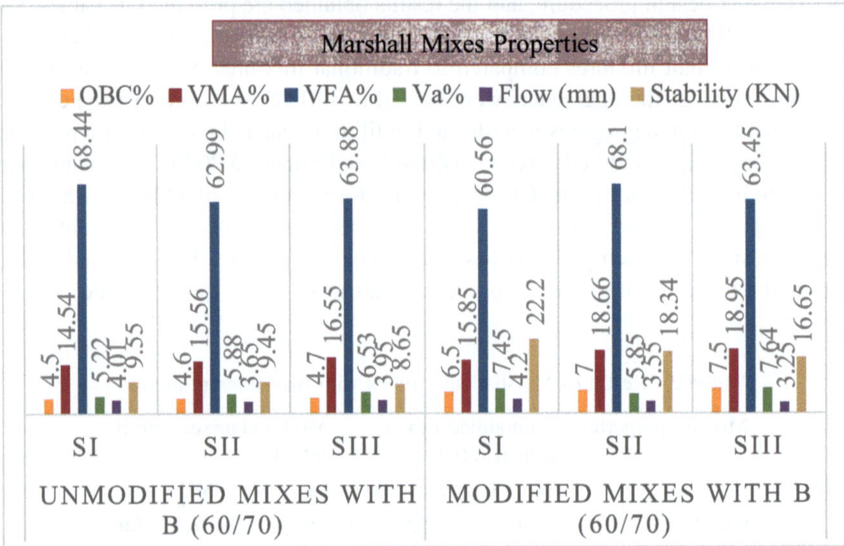

Fig. 2. Comparison between modified with CR and unmodified mixture under Marshall mix design.

after a curing period. The obtained results are shown in Tables 3 and 4 and illustrated in Figs. 2 and 3.

6 Synthesis and Examination of the Findings

The asphalt binder that was used in the Marshall method may not be sufficient to ensure that the selection of asphalt binders can satisfactorily meet the paving service temperatures experienced in Libya. The method of selecting the binder and the bituminous content in the mixture is still questionable. However, in this study, the conventional local specifications of selecting and blending the aggregate and binder that were used in the Marshall mix design did not perform well and thus yielded inadequate results. On the other hand, the Marshall mixes were designed using the CR, B60/70 and a proper aggregate gradation indicated that the Marshall method performed adequately. Asphalt-binder type, asphalt-binder content, aggregate properties, and gradation showed a

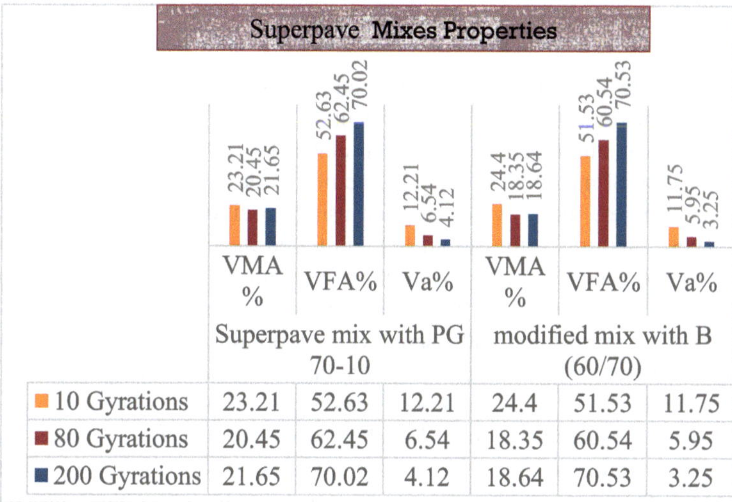

Fig. 3. Comparison between Superpave mix with PG 70-10 and modified mixture with CR and B60/70 under Superpave mix design.

significant effect on the performance of the mixes as expected. Tables 3 and 4 illustrate all the laboratory experiment results. Figures 2 and 3 show the behavior of the modifier-asphalt mixtures compared to traditional mixtures. Modifier-asphalt mixture and Superpave mixtures with new binder PG70-10 may be a proper choice for formulating the asphalt mixture under hot weather conditions such as in Libya. It can be seen from Fig. 2, Table 3 and Fig. 3, Table 4 that the volumetric property values of the Superpave mixes are properly better than that of Marshall mixtures. The flow values of modified mixes of the Marshall method are slightly less than that of unmodified mixes. The OBC, VMA, VFA, Va % and stability are more than that in unmodified mixes. This could be due to the differences in the mixture properties and may be related to the compaction effect. In Fig. 3, and Table 4, asphalt blends using Superpave with asphalt binder PG70-10 showed better properties than the modified blends using B 60/70 and CR at all stages of tests. The Superpave gyratory compactor rotates at a constant rate during the compaction, and this characteristic provides a better homogenous mixture of aggregate/PG 70-10 binder than the other mixture that contained a CR. This process simulates closely to the field compaction, reversing the compaction with Marshall hammer, which is unrealistic. The study clearly showed that the amount of air voids in the modified mixture with Marshall method is twice as much as that in the Superpave method. Although the number of air voids had a significant effect on the pavement performance, this parameter could not be relied on to predict the pavement performance. It is therefore important that air voids be carefully balanced during designing an asphalt mixture. Regardless of the type of mix design method used, it is important that the introduction of new generations of mix design methods and new additives are used to improve the properties of asphalt mixtures in Libya.

7 Conclusions and Recommendations

The use of modified mixtures or new generation methods for asphalt mix design can have a significant impact on the improvement of asphalt paving properties in Libya. However, the most important factors which affect the pavement performance are the environment, type of materials and the update of mix design methods. In this research, three asphalt concrete mixes were designed using modified, Marshall and Superpave mixtures. These mixtures were formulated and tested in accordance with the Marshall mix design method and the Superpave mix design method. The analysis was conducted to evaluate HMA volumetric properties such as air voids (Va), voids in mineral aggregate (VMA), voids filled with asphalt (FVA), optimum bitumen content, etc. The study has shown that in those in which the Marshall mixes were designed using CR and asphalt-binders B (60/70) and a proper aggregate gradation, the Marshall method performed better than the traditional Marshall mixes. The type of asphalt-binder, the asphalt-binder content, and the mixture properties, showed a significant effect on the performance of the mixes, as theoretically expected. On the other hand, the Superpave mixes were designed using PG 70-10, indicated that the Superpave mixture performed better than the modified mixture. Marshall modified mixtures and Superpave mixtures with new binder PG70-10 may be a proper choice for formulating the asphalt mixture under hot weather conditions such as in Libya. A virtual need to evaluate the mix-design methods on the hot weather is still mandatory.

Acknowledgements. The results presented in this paper are part of a Laboratory research on the several parameters that effect the asphalt concrete performance in hot arid weather. The authors would like to thank Bitumen laboratory (LCMB) staff at École de technologie Superior. It is with great joy and lightness of spirit that I offer my deepest regards to my supervisor Dr. Gabriel J. Assaf who has assisted and supported me in countless ways as I journeyed through the process of undertaking, creating, and completing this paper.

References

Almadwi, F.S., Assaf, G.J.: Performance testing of paving mixes for Libya's hot and arid conditions, using Marshall stability and SUPERPAVE gyratory compactor methods. In: International Congress and Exhibition Sustainable Civil Infrastructures: Innovative Infrastructure Geotechnology, pp. 313–323. Springer (2017)

Amorim, S.I.R., Pais, J.C., Vale, A.C., Minhoto, M.J.C.: A model for equivalent axle load factors. Int. J. Pavement Eng. **16**(10), 881–893 (2015)

Brule, B.: Polymer-modified asphalt cements used in the road construction industry: basic principles. Transp. Res. Rec. J. Transp. Res. Board **1535**, 48–53 (1996)

Charania, E., Cano, J.O., Schnormeier, R.H.: A twenty year study of asphalt-rubber pavements in the City of Phoenix, Arizona (1992)

Chesner, W.H., Collins, R.J., MacKay, M.H.: User guidelines for waste and by-product materials in pavement construction. FHWA report No. FHWA-RD-97-148 (1998)

El Khatib, A.K.: Development of a performance-based test method for quantification of cracking potential in asphalt pavement materials. Ph.D. dissertation in Civil and Environmental Engineering Department, University of Illinois at Urbana-Champaign (2016)

Haixu, Z., Yongliang, H., Haibo, X.: Application and error of modified asphalt technology in municipal road engineering. Urban Roads Bridg Flood Control **8**, 43 (2012)

Heitzman, M.: Design and construction of asphalt paving materials with crumb rubber modifier. Transp. Res. Rec. **1339**, 1–8 (1992)

Kaya, A.: Potential use of tire rubber and ebonite in asphalt: synthesis study: executive summary. Joint Highway Research Project No.: C-36-50M between Engineering Experiment Station Purdue University. In Cooperation with the Indiana Department Transportation and the U.S Department of Transportation Federal Highway administration (1992)

Park, S.W., Kim, Y.R., Schapery, R.A.: A viscoelastic continuum damage model and its application to uniaxial behavior of asphalt concrete. Mech. Mater. **24**(4), 241–255 (1996)

Roy, N., Veeraragavan, A., Jain, P.K., Krishnan, J.M.: A re-look at the asphalt mixture performance test protocols and computational algorithms. J. Test. Eval. **41**(5), 729–744 (2013)

Salini, R.B.: Utilização de borracha reciclada de pneus em misturas asfálticas (2000)

Zhu, J., Birgisson, B., Kringos, N.: Polymer modification of bitumen: advances and challenges. Eur. Polym. J. **54**, 18–38 (2014)

Zoorob, S.E., Suparma, L.B.: Laboratory design and investigation of the properties of continuously graded Asphaltic concrete containing recycled plastics aggregate replacement (Plastiphalt). Cement Concr. Compos. **22**(4), 233–242 (2000)

Evaluating Falling Weight Deflectometer Back-Calculation Software for Aircraft Pavement Strength Rating

Greg White[1(✉)] and Andrew Barbeler[2]

[1] Airport Pavement Research Program, University of the Sunshine Coast, Sippy Downs, QLD, Australia
gwhite2@usc.edu.au

[2] University of the Sunshine Coast, Sippy Downs, QLD, Australia

Abstract. In recent years, software has been developed to calculate aircraft pavement strength ratings from modulus values that are back-calculated from deflections measured by a falling weight deflectometer. Deflection data from a uniform section of consistent runway pavement was analysed for variability to determine the effect of deflection variability on back-calculated modulus and the resulting pavement strength rating. The drop force of the falling weight had a significant influence on the calculated pavement strength, with 50 kN drop force rating the pavement approximately 35% lower than the 100 kN drop force. The number of layers in the nominated pavement structure also had a significant influence on the back-calculated modulus values, but not the resulting pavement strength rating. Compared to traditional design-based evaluation using intrusively measured layer thicknesses and laboratory material characterisation, the falling weight deflection back-calculation underestimated the pavement strength. Consequently, it is recommended that airports do not set their PCN based on back-calculated FWD data unless also supported by intrusive geotechnical investigation and a design-based evaluation.

1 Introduction

With airports getting busier and funding for upgrade works limited, non-destructive testing (NDT) which is rapid to perform and relatively economical to procure, has gained popularity. NDT is particularly useful for busy airport runways that can not readily be closed for days of traditional intrusive pavement investigation (White 2017). The falling weight deflectometer (FWD) is arguably the most common NDT device used for the structural evaluation of existing airport pavement infrastructure (Celaya and Nazarian 2014).

FWDs have been in use since the 1980s. These generally trailer-mounted NDT devices apply a dynamic load while the trailer is stationary. The load can be varied and deflections are estimated from surface strain accelerations measured by geophones placed at various distances from the load application (TRB 2008). This results in a deflection profile or bowl that is a function of the composition and condition of the pavement (Vuong 1989). In recent years softwares have been developed for the estimation of layer modulus values from deflection data, commonly known as

S. El-Badawy and R. Abd El-Hakim (Eds.): GeoMEast 2018, SUCI, pp. 64–83, 2019.
https://doi.org/10.1007/978-3-030-01908-2_6

back-analysis, and the resulting modulus values are used to calculate a pavement strength rating based on a nominated design aircraft (Dynatest 2017).

This paper investigates the use of FWD results for aircraft pavement strength rating. A consistent runway pavement structure was surveyed by FWD and a back-calculation software was used to estimate layer modulus values and a pavement strength rating at each FWD test point. The variability of the deflection, modulus and strength rating results were evaluated to determine the risk of inappropriate strength rating compared to more traditional design-based methods.

2 Background

2.1 Airport Pavement Strength Rating System

Airport pavements are usually either flexible or rigid structures, although composite pavements are also used in some circumstances. Flexible pavements usually comprise significant thickness of crushed rock base and sub-base with a thin (50–100 mm) or thick (100–250 mm) asphalt surface, although some regional airports use a bituminous sprayed seal wearing surface.

Airport pavements are subject to an international strength rating system, known as ACN–PCN (ICAO 1983). The system is similar to road axle load limits, but reflects the higher level of deviation between small aircraft and large aircraft, meaning every airport has its own Pavement Classification Number (PCN), which represents the upper limit of Aircraft Classification number (ACN) allowed for unrestricted aircraft operations.

Traditionally an airport's PCN was determined by reverse engineering the existing pavement structure, including layer thicknesses and materials, as well as the subgrade CBR, to determine the acceptability of a particular aircraft operating at a particular mass. The PCN was generally set to the highest ACN of all the aircraft considered to be acceptable for the particular pavement. This approach remains the normal method for PCN assignment in many countries (FAA 2011). However, as detailed below, software is now available to automatically calculate a PCN, primarily from FWD deflection data.

2.2 Falling Weight Deflectometer

There are at least four manufacturers of FWD devices, although Dynatest is the most commonly encountered brand. All FWDs apply a vertically dynamic load, generally 20–65 ms in duration, onto a 300–450 mm diameter load plate (Ameri et al. 2009). The load generally ranges from 7 kN to 150 kN for the standard FWD, although a light version (LWD) applies 1–15 kN and various heavy versions (HWD) apply up to 250 kN (FAA 2011). Deflections are estimated from geophone measurements of surface acceleration, that can be spaced as required, usually anywhere from 0–2400 mm from the centre of the loading plate. Standard distances for geophone locations are 0, 300 600, 900, 1200, 1500 and 1800 mm from the centre of the load plate. However, different countries adopt different spacings. For example, Australia specifies a minimum of seven geophones, with three located 0, 200 and 900 mm from the load plate centre, with the location of the other geophones optional (Austroads 2011). This approach reflects the focus on the deflections at 0, 200 and 900 mm for the characterisation of the deflection bowl, using the generic term 'D-n' to represent the deflection at the geophone located n mm from the centre of the load pate.

Deflection bowls are generally characterised by three parameters (Austroads 2011):

- Maximum deflection (D-0). A general indicator of pavement stiffness and response.
- Curvature (D-0 minus D-200). An indicator of the upper base course and surface layer stiffness.
- D-900. An indicator of the subgrade support condition.

It is important to understand that FWD deflections are likely to change with seasonal variations. As the surface temperature increases, the bituminous binder stiffness reduces and pavement deflections increase (Jin et al. 1994). Similarly, as the subgrade and granular material moisture content decreases, the material stiffness increases and deflections reduce (Deblois et al. 2010). Consequently, deflections increase and back calculated modulus values decrease during the hotter and wetter seasons. For example, seasonal subgrade back-calculated modulus results fluctuated by up to 20% and the spatial variability fluctuated due to the non-uniform distribution of moisture content increase, when tested every month throughout a year (Hossain et al. 2000).

2.3 Modulus Back-Calculation

A number of softwares are now available for the back-calculation of layer modulus values from deflection bowl data. Example softwares include Modulus, Elmod, Evercalc (Ameri et al. 2009), DAPS (Bandara et al. 2002), Elsedef, Illi-Back (Kang 1998) and EfromD (Vuong 1989). Generally, the software requires pavement layer thicknesses to be nominated, as well as modulus seed values. Although attractive to the designer, absolute back calculated modulus values should be used with caution (Gendreau and Soriano 1998) due to the high degree of variance between predicted and observed pavement performance (Zaniewski 1991).

Due to the prevalence of the Dynatest FWD, Dynatest's ELMOD (Dynatest 2017) is arguably the most common back analysis software used by practitioners. The latest version, ELMOD 6 also includes PCN determination from deflection bowl data.

Following layer modulus estimation, the software uses nominated aircraft type and frequency to determine the most damaging aircraft for each nominated pavement layer. The ACN of the most damaging aircraft at the 'just acceptable' aircraft mass is reported as the allowable pavement PCN. The allowable mass may be higher than the published maximum mass of the aircraft in order to 'just fail' the pavement by the end of the design life. The pavement damage caused by each aircraft is based on an allowable stress or strain level for each pavement layer. This approach does not take into account superposition or the impact of multiple aircraft types, relies upon the allowable stress/strain determined for various pavement materials, and relies upon accurate pavement layer thicknesses.

In ELMOD 6, every test location has a PCN value reported. The PCN values vary and a protocol is required for determining a characteristic PCN from the hundreds or thousands of test points. Typical protocols include selecting the 5%-ile or 10%-ile value such that the majority of the calculated PCN values exceed the selected value. However, it is not known how these values will typically compare to a design-based PCN evaluation using measured pavement material properties and layer thicknesses from intrusive geotechnical testing.

3 Methods and Results

As part of a major pavement upgrade project a FWD survey was conducted of the main runway at Dubbo Airport, located in central New South Wales, Australia. The FWD survey included five drops, at increasing drop force, every 10 m along the runway, on the centreline, as well as 3 m and 6 m on both sides of the centreline. The survey was completed over two nights in September 2016 and the 50 kN (drop three) and 100 kN (drop five) results were recorded and analysed (Figs. 1 and 2). The 1,000 m length of runway between 400 m and 1,400 m from the western end of the runway was determined to be consistent. This was confirmed by intrusive testing to physically measure the layer thicknesses and sample materials for base, sub-base and subgrade testing (Table 1).

3.1 Methods

The FWD results from the consistent portion of the runway length were analysed for variability and then for back-calculation to estimate layer modulus using ELMOD 6. ELMOD 6 was also used to estimate PCN values at each test location, based on the Embraer E190 aircraft as the critical aircraft. The analysis considered two pavement structures, one with the bituminous surface modelled as a separate layer and one with the bituminous surface thickness incorporated into the base course thickness (Table 2). The results were statistically analysed for variability and trends to determine the influence of the FWD drop force and number of layers. The influence of key FWD responses on modulus and PCN values was determined by linear regression models.

3.2 Results

The maximum deflection, at the point of load application (D-0) and the curvature (CV), defined as the D-0 less the deflection 200 mm from point of load application (D-200) are shown in Fig. 1 (50 kN drop force) and Fig. 2 (100 kN drop force). Summary statistics for the three-layer model back-calculated modulus values are in Tables 3 and 4, for 50 kN and 100 kN FWD drop forces, respectively. Similarly, the four-layer model modulus value statistics are in Tables 5 and 6, while the subsequent PCN values are in Table 7.

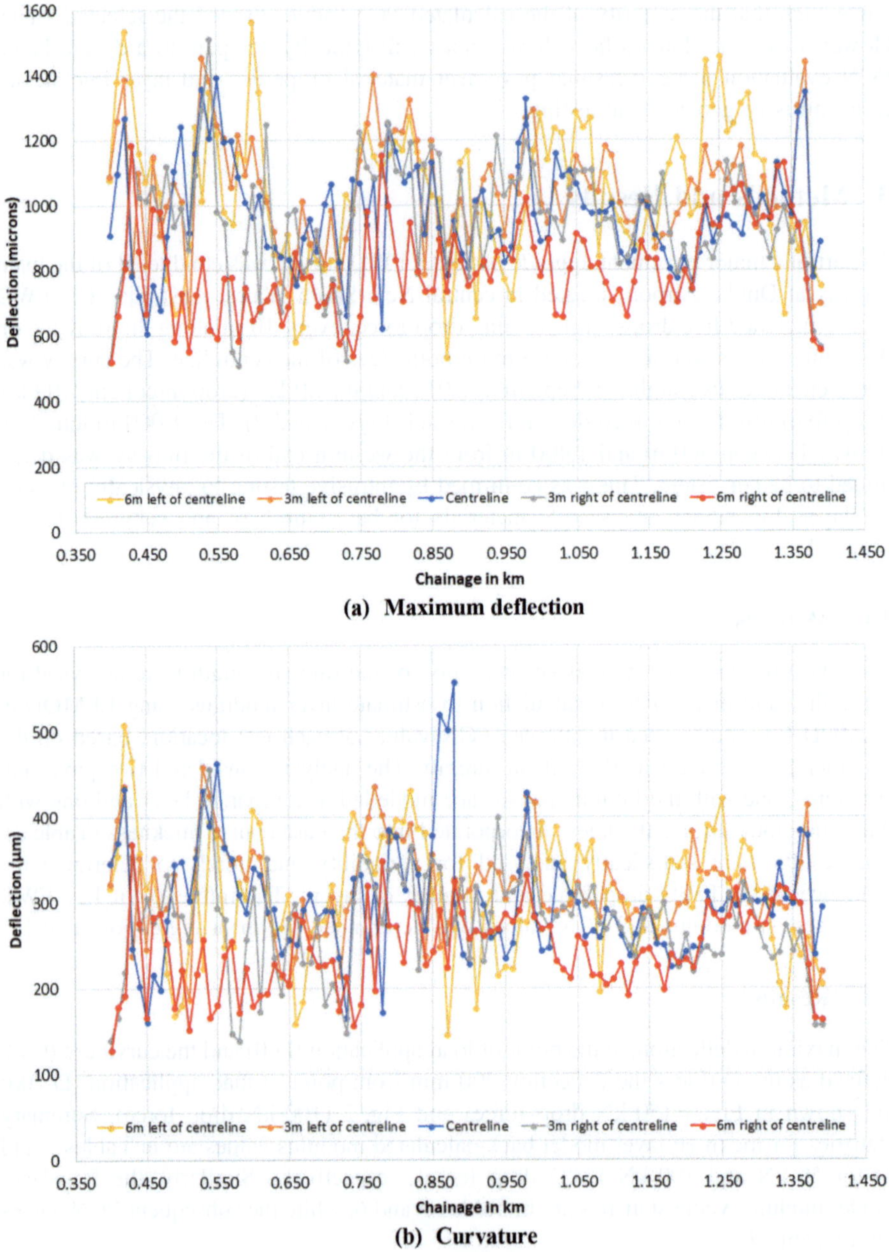

Fig. 1. 50 kN FWD D-0 and CV

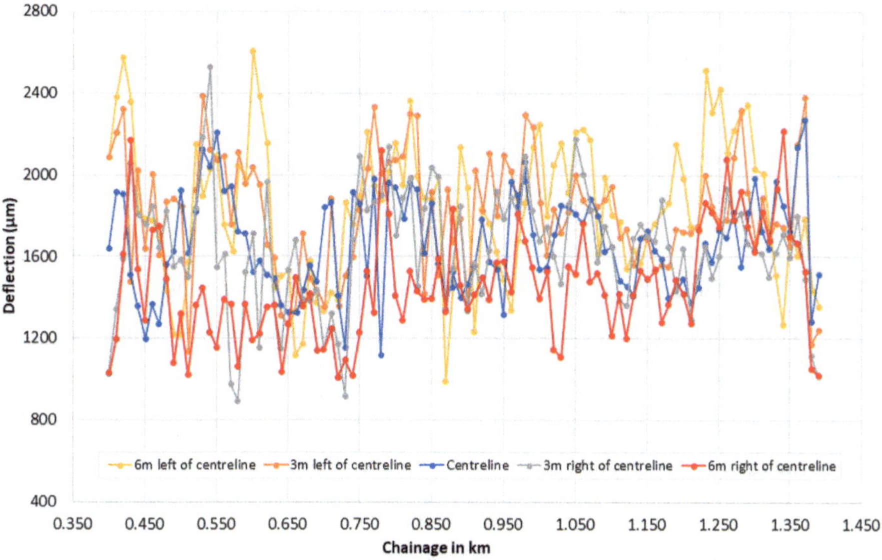

(a) Maximum deflection

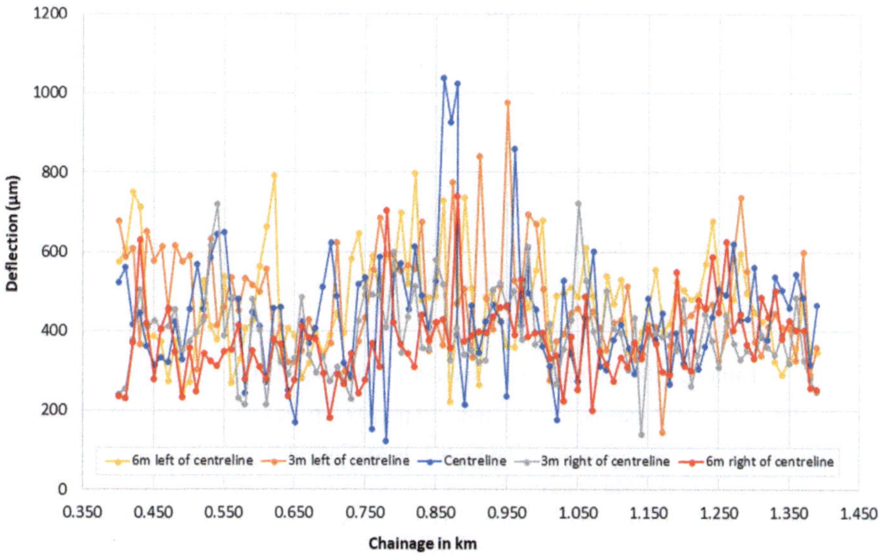

(b) Curvature

Fig. 2. 100 kN FWD D-0 and CV

Table 1. Pavement structure from intrusive testing

Pavement layer	Depth	Material
Surface	0–40 mm	Various sprayed seals
Base	160–200 mm	Natural gravel
Sub-base	200–400 mm	Natural gravel
Subgrade	400 mm and below	CBR 5% clay

Table 2. Three and four layer pavement models for ELMOD 6

Pavement layer	Three layer model thickness	Four layer model thickness
Asphalt surface	Not applicable	40 mm
Base	200 mm	160 mm
Sub-base	200 mm	200 mm
Subgrade	Infinite	Infinite

Table 3. 50 kN three layer modulus and subgrade CBR statistics

Statistic	Modulus (MPa)		CBR (%)
	Base	Sub-base	Subgrade
Minimum	203	23	2
Q1	454	45	9
Average	541	82	16
Q3	608	80	18
Maximum	1,135	2,658	63
Range	932	2,635	61
Std. Dev.	136	156	11
CoV	25%	190%	68%

Table 4. 100 kN three layer modulus and subgrade CBR statistics

Statistic	Modulus (MPa)		CBR (%)
	Base	Sub-base	Subgrade
Minimum	236	25	1
Q1	610	47	8
Average	740	97	14
Q3	838	97	16
Maximum	1,703	4,101	58
Range	1,467	4,075	57
Std. Dev.	194	254	9
CoV	26%	261%	63%

Table 5. 50 kN four layer modulus and subgrade CBR statistics

Statistic	Modulus			CBR (%)
	Surface	Base	Sub-base	Subgrade
Minimum	72	53	27	1
Q1	9,924	138	78	7
Average	13,251	181	162	11
Q3	15,674	209	168	12
Maximum	34,319	764	3,943	61
Range	34,247	711	3,916	61
Std. Dev.	5,086	70	258	7
CoV	38%	39%	159%	67%

Table 6. 100 kN four layer modulus and subgrade CBR statistics

Statistic	Modulus			CBR (%)
	Surface	Base	Sub-base	Subgrade
Minimum	61	48	23	1
Q1	10,833	148	75	7
Average	20,942	250	173	11
Q3	29,155	306	177	12
Maximum	77,122	1,258	4,816	57
Range	77,061	1,210	4,793	56
Std. Dev.	13,697	154	363	8
CoV	65%	62%	210%	69%

Table 7. Four layer Pavement Classification Numbers statistics

Statistic	Pavement Classification Number	
	50 kN	100 kN
Minimum	2	2
Q1	4	4
Average	5	6
Q3	6	7
Maximum	26	17
Range	24	14
Std. Dev.	2	3
CoV	45%	46%

4 Discussion

4.1 Deflection, Modulus and PCN Variability

The measured deflections clearly reduced with increasing distance from the point of load application and increased with applied load magnitude (Fig. 3). Within a section of generally uniform pavement, the variability in the FWD responses was moderate, with a CoV ranging from 16% to 21%. The variability was no different for the 50 kN and 100 kN FWD drop forces and D-900 was slightly less variable than the D-0 and D-200 results.

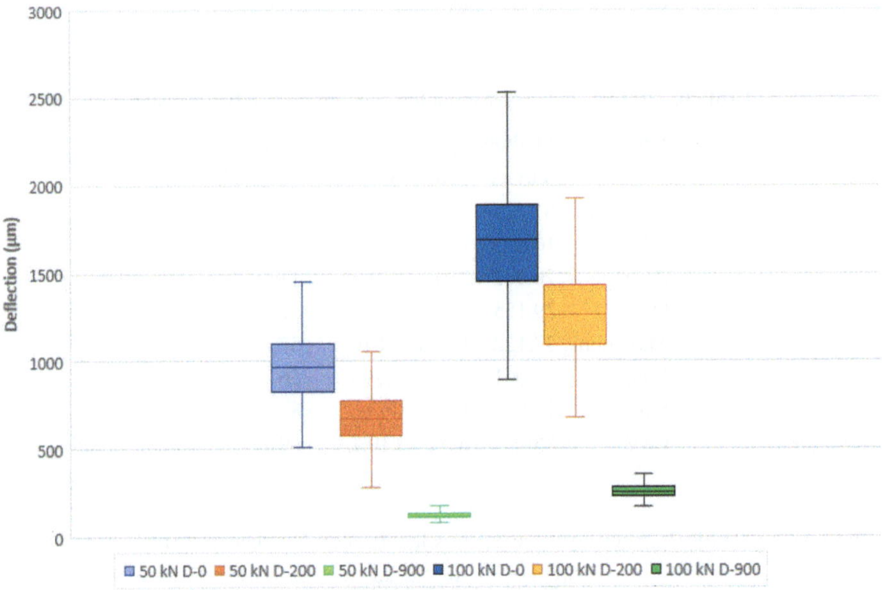

Fig. 3. FWD deflections

The surface layer modulus calculated by ELMOD 6 for the four-layer model was high, ranging from 72 MPa to 34,319 MPa, with an average of 13,251 MPa (Table 5). The high surface layer modulus of the four-layer model was offset by a lower base layer modulus, compared to the three-layer model values (Fig. 4). In contrast, the three-layer system sub-base modulus was lower than for the four-layer model, likely to compensate for the low four-layer model base course modulus forced by the high surface modulus. However, the three- and four-layer model sub-base modulus values are much closer than the base modulus values. Overall, the modulus values were more variable than the deflections, with CoVs ranging from 25% to 261%. The 100 kN modulus values were more variable than the 50 kN modulus values and the four-layer model modulus values were more variable than the three-layer values (Table 8).

The ELMOD 6 calculated subgrade modulus values were converted to subgrade CBR (modulus divided by 10) and the results were more variable than the deflections,

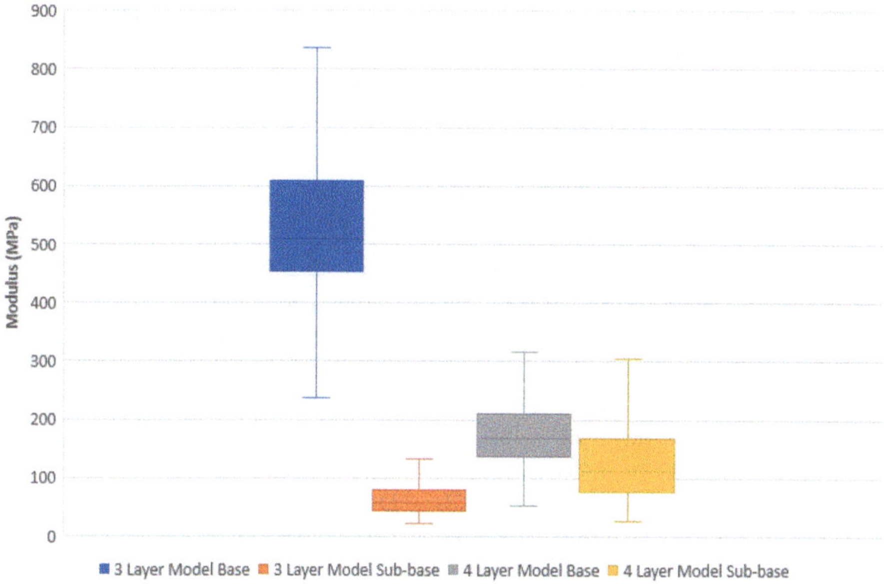

(a) 50 kN drop force

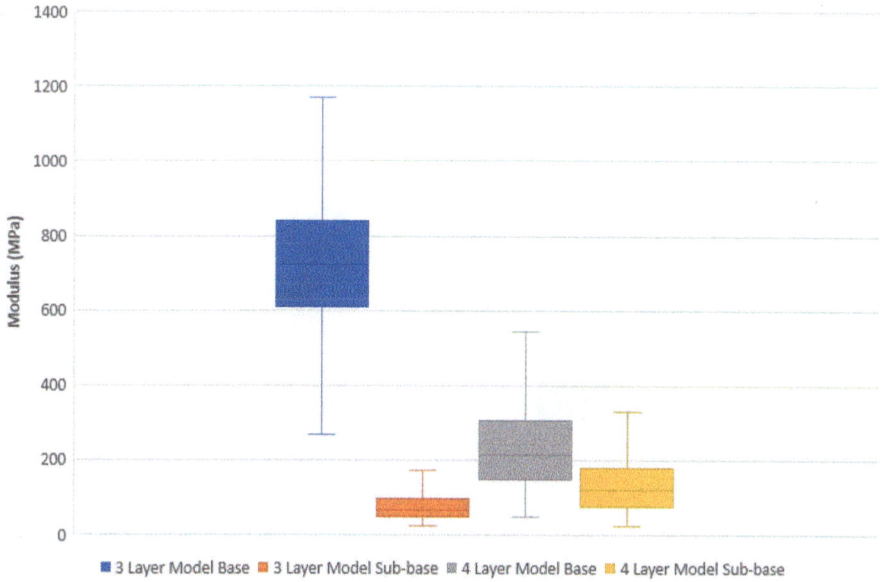

(b) 100 kN drop force

Fig. 4. ELMOD 6 calculated base and sub-base modulus values

with CoV values from 63% to 69% (Tables 3, 4, 5 and 6). Ranging from approximately CBR 1 to CBR 60, for both FWD drop forces and both the three- and four-layer models, with even the central 50% of the results ranging from approximately CBR 8%

to CBR 14%. This is a significant portion of the realistic range of design subgrade CBR values, which is generally accepted as CBR 3-15% (AAA 2017). The three-layer model returned higher subgrade CBR values than the four-layer model and the FWD drop force was not significant for calculated subgrade CBR (Fig. 5).

Table 8. Deflection statistics

Statistic	50 kN			100 kN		
	D-0	D-200	D-900	D-0	D-200	D-900
Minimum	508	283	79	894	425	170
Q1	827	577	109	1,456	1,089	226
Average	965	675	123	1,685	1,255	256
Q3	1,099	776	136	1,890	1,428	284
Maximum	1,562	1,155	210	2,607	2,044	425
Range	1,054	872	130	1,713	1,619	255
Std. Dev.	197	141	21	319	245	42
CoV	20%	21%	17%	19%	20%	16%

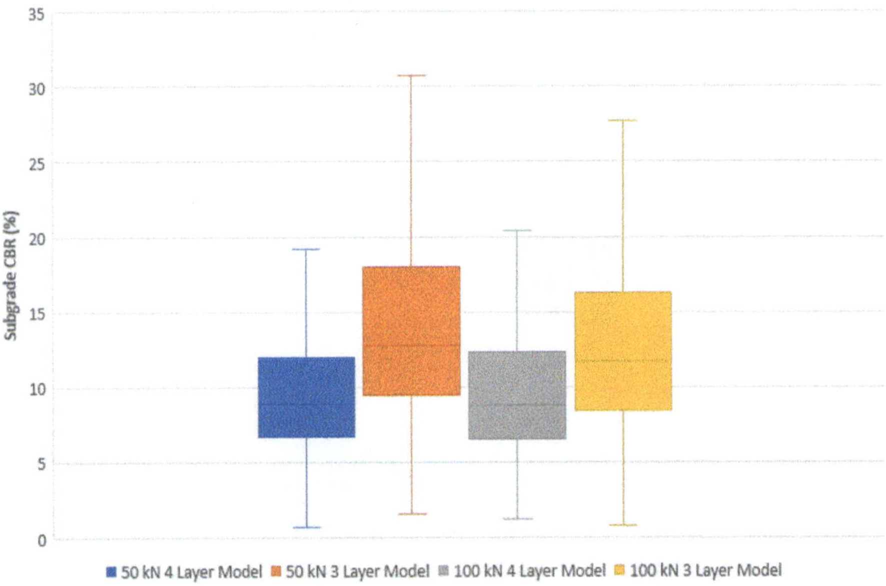

Fig. 5. ELMOD 6 calculated subgrade CBR values

4.2 Comparison of Three or Four Layer Pavement Model

As discussed above the four-layer ELMOD 6 pavement model assigned a generally high modulus to the 40 mm thick bituminous surface layer. In fact, the modulus values assigned averaged over 13,000 MPa for the 50 kN FWD drop force and over

20,000 MPa for the 100 kN testing. These values significantly exceed dense graded asphalt modulus values for Australian airports, which generally do not exceed 6,000 MPa, even for stiff polymer modified binders.

As a result of the high surface layer modulus, the base course layer modulus was significantly higher for the three-layer model than for the four-layer model. The relationship between the three-layer and four-layer model base course modulus was more consistent ($R^2 = 0.35$) for the 50 kN drop force than for the 100 kN drop force ($R^2 = 0.02$), as shown in Fig. 6.

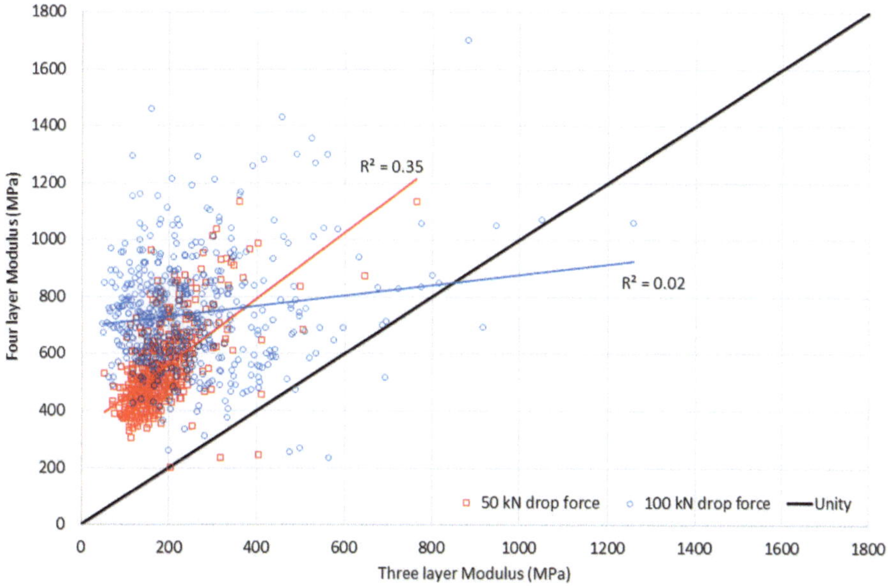

Fig. 6. ELMOD 6 three and four layer base modulus

The relationships were more consistent for the sub-base modulus, with the 100 kN drop force showing better agreement between the three- and four-layer models, than the 50 kN drop force (Fig. 7). The better agreement associated with the sub-base likely reflects the compensation for the high surface layer modulus mostly occurring in the base course layer modulus values. However, as discussed previously, the three-layer model sub-base modulus values were consistently around 35% lower than the four-layer model values (Fig. 7). Similarly, the subgrade CBRs were, on average, not significantly different for the three- and four-layer models (Fig. 8) but the agreement was more variable than the sub-base modulus values, with linear regression R^2 values of 0.36 (50 kN drop force) and 0.31 (100 kN drop force).

Finally, the agreement between the PCN values calculated by ELMOD 6 from the three-layer and four-layer models was better (Fig. 9). On average, the two models returned similar PCN values and the agreement was much higher than for the subgrade

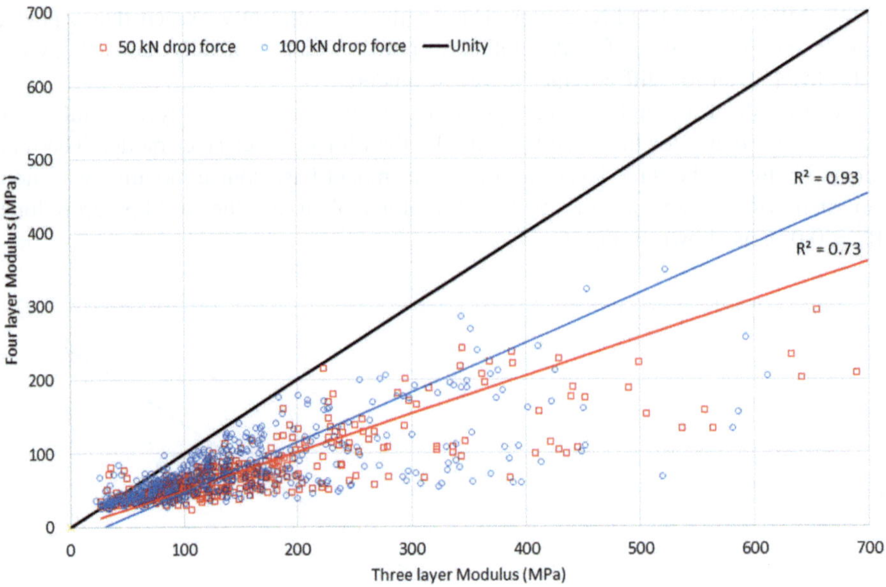

Fig. 7. ELMOD 6 three and four layer sub-base modulus

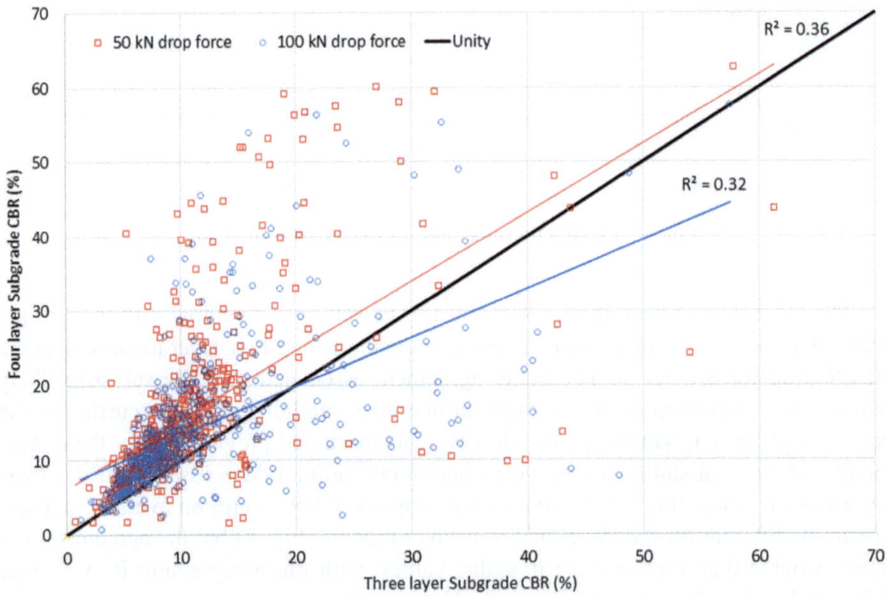

Fig. 8. ELMOD 6 three and four layer subgrade CBR

CBR values, with $R^2 = 0.82$. Further detail on the effect of layer modulus and subgrade CBR on ELMOD 6 calculated PCN is provided later.

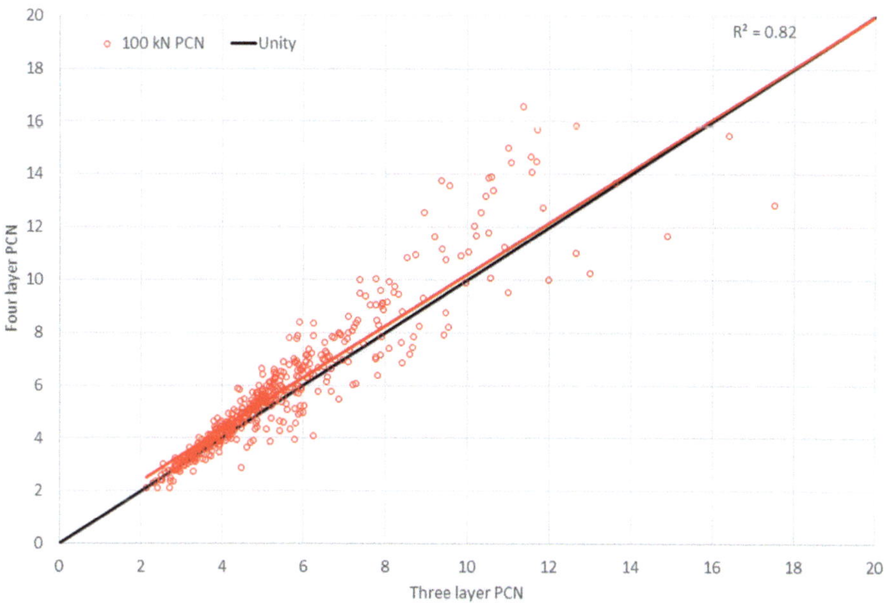

Fig. 9. ELMOD 6 three and four layer pavement PCN

4.3 Comparison of 50 kN and 100 kN FWD Results

For the three-layer model, there was reasonable consistence between the base course modulus values for the 50 kN and 100 kN drop force results (Fig. 10). However, Fig. 10 also indicates that the consistence was minimal for the four-layer model, which likely reflects the different influence of the four-layer model surface modulus on the resulting base course modulus. In contrast, the sub-base modulus was consistent for both the three-layer and four-layer models (Fig. 11). On average, the sub-base modulus for the 100 kN FWD drop force deflection was approximately 20% higher than for the 50 kN FWD drop force, reflecting the stress-dependence of the granular base course materials. In contrast, the subgrade CBR was, on average, 30% lower for the 100 kN FWD drop force than for the 50 kN FWD drop force (Fig. 12). Because the PCN values were only calculated for the 100 kN FWD drop force results, no comparison was made.

4.4 Sensitivity of Layer Modulus to Deflection Changes

Linear regressions were calculated with D-0, D-200 and D-900 as independent variables and the modulus of the various pavement layers and the subgrade CBR as the dependent variables. This allowed the sensitivity of the layer modulus and subgrade CBR to the primary deflection bowl characteristics to be calculated (Table 9). A constant was included in the regression models. In all cases, all three deflection bowl characteristics were statistically significant for all modulus values and subgrade CBR values, with p-values much less than 0.05 (Table 10).

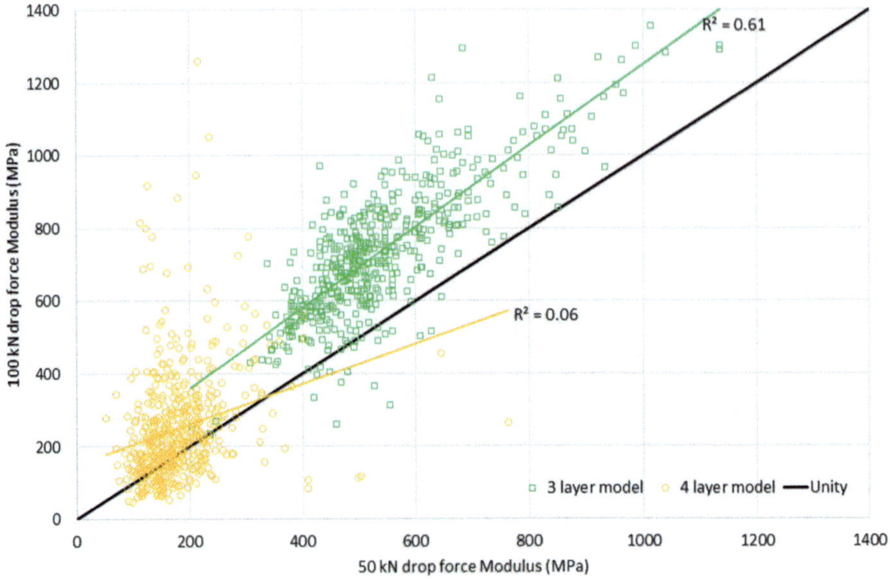

Fig. 10. ELMOD 6 50 kN and 100 kN base modulus

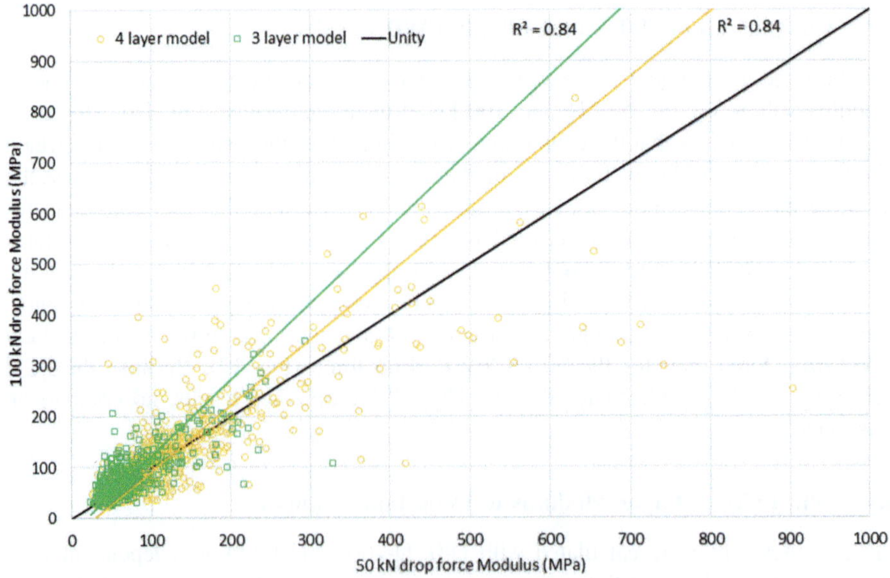

Fig. 11. ELMOD 6 50 kN and 100 kN sub-base modulus

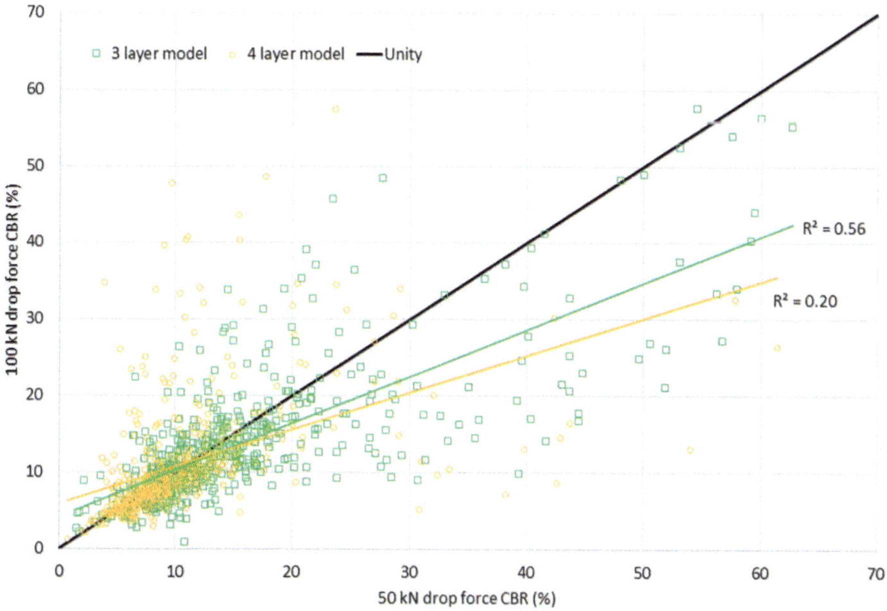

Fig. 12. ELMOD 6 50 kN and 100 kN pavement PCN

Table 9. Modulus regression model factor coefficients

Regression model	Constant	D-0	D-200	D-900
3 layer, base modulus	1,389	−1.18	0.99	0.37
3 layer, sub-base modulus	152	1.06	−1.8	1.7
3 layer, subgrade CBR	48.5	−0.014	0.01	−0.09
4 layer, surface modulus	51,824	−81.8	93.4	−39.9
4 layer, base modulus	286	0.26	−0.53	0.71
4 layer, sub-base modulus	422	1.37	−2.53	2.38
4 layer, subgrade CBR	31.2	−0.015	0.02	−0.1

Table 10. Modulus regression model p-values

Regression model	Constant	D-0	D-200	D-900
3 layer, base modulus	0.00	0.00	0.00	0.00
3 layer, sub-base modulus	0.00	0.00	0.00	0.00
3 layer, subgrade CBR	0.00	0.00	0.01	0.00
4 layer, surface modulus	0.00	0.00	0.00	0.00
4 layer, base modulus	0.00	0.00	0.00	0.00
4 layer, sub-base modulus	0.00	0.00	0.00	0.00
4 layer, subgrade CBR	0.00	0.00	0.00	0.00

4.5 Sensitivity of PCN to Layer Modulus

Linear regression models for ELMOD 6 calculated PCN values indicated that all deflection bowl characteristics were statistically significant except for the influence of D-0 on the three-layer PCN (Table 11). The simple regression models were reasonable predictors, with R^2 values of 0.62 and 0.73, for the three-layer and four-layer models, respectively.

Table 11. PCN regression model factor coefficients

Regression model	Constant	D-0	D-200	D-900
3 layer, pavement PCN	13.5	−0.0003	−0.0085	0.012
4 layer, pavement PCN	15.5	−0.0019	−0.0081	0.014

Using D-0, D-200 and D-900 values one and two standard deviations either side of the mean, ELMOD 6 inferred PCN values were calculated using the factor coefficients in Table 11 as the average rate of change of PCN due to a change in the deflection characteristic. The resulting PCN values ranged from 12.1 to 18.9, for two standard deviations below and above the mean of each deflection characteristic, respectively (Table 12). This range includes approximately 85% of the all the ELMOD 6 calculated PCN values but there is no basis for selecting one value over another. Consequently, the PCN values calculated by ELMOD 6 from FWD deflection data should be considered to be approximate only.

Table 12. Regression PCN values for different deflection values

Deflection characteristic	−2 SD	−1 SD	Mean	+1 SD	−2 SD
D-0	1046	1365	1685	2004	2324
D-200	764	1010	1255	1500	1746
D-900	172	214	256	298	340
Resulting PCN	12.1	13.8	15.5	17.2	18.9

4.6 Comparison of FWD PCN to Pavement Design PCN

The efficacy of PCN values calculated by ELMOD 6 from FWD deflections was determined by comparison to more traditional design-based PCN values. The design-based PCN values were determined by analyzing the adopted pavement structure (Table 2) in the flexible pavement design software APSDS (Wardle and Rodway 2010) under E190 aircraft loadings. The E190 aircraft mass was varied until the pavement was modelled to just fail under various parameter values (Table 13) and the pavement PCN was determined to be equal to the E190 ACN at that aircraft mass (Table 14).

The design-based PCN values ranged from 4.6 to 25.7, with the 10,000 passes and CBR 5% value being PCN 12.1. When compared to the PCN values calculated by ELMOD 6, the design-based PCN values correlated with the 44–100%-ile PCN values

Table 13. Design-based PCN parametric factors

Factor	Factor values		
Subgrade CBR	3	5	7
Aircraft passes	1,000	10,000	100,000

Table 14. Design-based PCN values

E190 aircraft passes	Subgrade CBR		
	3	5	7
1,000	7.9	16.1	25.7
10,000	6.2	12.1	19.7
100,000	4.6	9.6	15.4

(Fig. 13). The 10,000 aircraft coverage and subgrade CBR 5% PCN of 12.1 was associated with a 96%-ile (four-layer model) and 98%-ile (three-layer model) PCN values calculated from ELMOD 6. Consequently, the common adoption of the 10%-ile value (PCN 3.2) is significantly lower than determined by design-based analysis. It follows that airports that set their PCN based of FWD and ELMOD 6 data analysis are likely to under-rate the true structural capacity of the pavement.

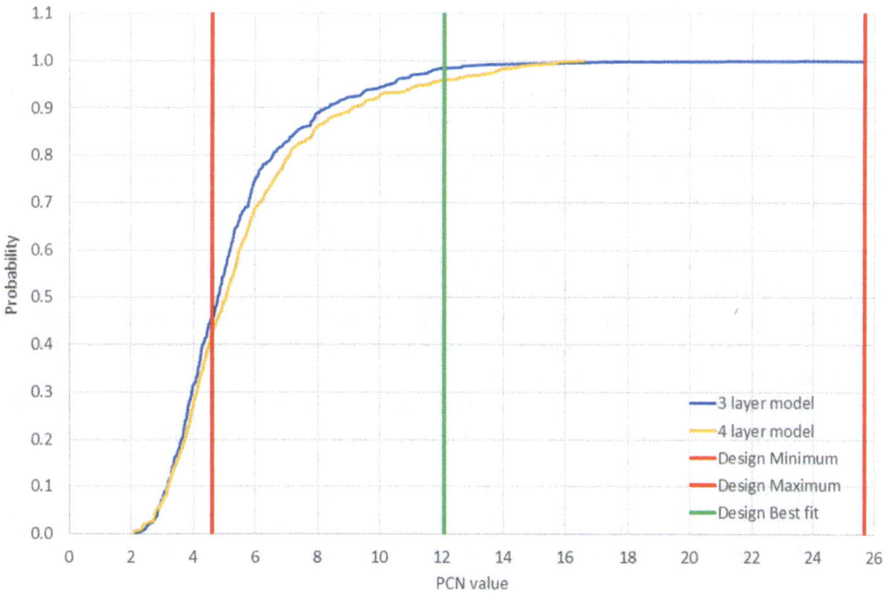

Fig. 13. Design-based and ELMOD 6 PCN value probability

5 Conclusions

The FWD is a useful NDT device for the rapid collection of data indicating pavement response to aircraft loading. This application is informative for determining sections of uniform pavement strength and the relative strength of each uniform area. However, extending the application to calculation of absolute layer modulus values and reliance on those modulus values for rating and publishing the strength of aircraft pavements is not recommended. The reliance of the ELMOD 6 software on accurate pavement layer thicknesses and the high variability in modulus values results in PCN values that span all reasonable PCNs. This tended to under-rate the true strength of the pavement determined by design-based structural evaluation using intrusive geotechnical investigation and laboratory material testing. Consequently, it is recommended that airports do not set their PCN based on back-calculated FWD data unless also supported by intrusive geotechnical investigation and a design-based evaluation. However, further research is required to extend these findings to different airport pavement structures.

Acknowledgements. The FWD and geotechnical investigation data provided by Dubbo Regional Council, ARRB, and Macquarie Geotech is greatly appreciated and gratefully acknowledged.

References

AAA: Airfield pavement essentials, Airport Practice Note 12, Australian Airports Association, Canberra, Australia, April 2017. https://airports.asn.au/public/member-centre/resources. Accessed 30 Nov 2017

Ameri, M., Yavari, N., Scullion, T.: Comparison of static and dynamic backcalculation of flexible pavement layer moduli, using four software programs. Asian J. Appl. Sci. **2**(3), 197–210 (2009)

Austroads: Pavement deflection measurement with a falling weight deflectometer. Austroads Test Method AG:AM/T006, 31 March 2011

Bandara, N., Rowe, G.M., Sharrock, M.J., Nickerson, C.R.: Seasonal variation of subgrade modulus in different subgrade soils for pavement rehabilitation for non freeze-thaw cycles. Appl. Adv. Technol. Transp. 473–480 (2002)

Celaya, M., Nazarian, S.: Field evaluation of NDT devices for delamination detection of HMA airport pavements. In: FAA Worldwide Airport Technology Transfer Conference, Galloway, New Jersey, USA, 5–7 August 2014

Deblois, K., Bilodeau, J.-P., Dore, G.: Use of falling weight deflectometer time history data for the analysis of seasonal variation in pavement response. Can. J. Civ. Eng. **37**, 1224–1231 (2010)

Dynatest: ELMOD 6 Quick Start Manual, Dynatest International (2017)

FAA: Use of nondestructive testing in pavement evaluations. Advisory Circular 150/5370-11B, Federal Aviation Administration, Washington, District of Columbia, USA, 30 September 2011

Gendreau, M., Soriano, P.: Airport pavement management systems: an appraisal of existing methodologies. Transp. Res. **32**(3), 197–214 (1998)

Hossain, M., Romanoschi, S., Gisi, A.J.: Seasonal and spatial variation of subgrade response. In: GeoDenver, Denver, Colorado, USA, 5–8 August 2000, pp. 150–166 (2000)

ICAO: Aerodrome Design Manager: Part 3, ICAO 9157, 2nd edn. International Civil Aviation Organization, Montreal, Quebec, Canada, January 1983

Jin, M.S., Lee, W.K., Kovacs, W.D.: Seasonal variation of resilient modulus of subgrade soils. J. Transp. Eng. **120**(4), 603–616 (1994)

Kang, Y.V.: Multifrequency back-calculation of pavement layer moduli. J. Transp. Eng. **124**(1), 73–81 (1998)

TRB: Falling Weight Deflectometer Usage, NCHRP Synthesis 381, Transportation Research Board, Washington, District of Columbia, USA (2008)

Vuong, B.: A new linear elastic back-calculation model for back-calculating layer moduli at fixed Poisson's ratio. Aust. Road Res. **19**(1), 17–28 (1989)

Wardle, L.J., Rodway, B.: Advanced design of flexible aircraft pavements. In: Proceedings of the 24th ARRB Conference, Melbourne, Victoria, Australia, 12–15 October 2010

White, G.: Expedient runway upgrade technologies. In: Proceedings of the 10th International Conference on the Bearing Capacity of Roads, Railways and Airfields, Athens, Greece, 28–30 June 2017

Zaniewski, J.: Unified Methodology for Airport Pavement Analysis and Design, Technical Report DOT/FAA/RD-91/15,I, Federal Aviation Administration, Washington, District of Columbia, USA, June 1991

Estimating VOC Emissions from Asphalt Pavement

Peng Cui[1], Qingbin Cui[1(✉)], and Harold Green[2]

[1] University of Maryland, College Park, MD, USA
cui@umd.edu
[2] Chamberlain Contractors Inc., Laurel, MD, USA

Abstract. Asphalt pavement is widely used in the US. The Volatile Organic Compounds (VOCs) volatized and emitted during its life cycle phase, including asphalt production, transportation, pavement paving, operation, as well as the demolition and reclaim, have a great effect on human health and the environment. Many studies have focused on the calculation, collecting, testing, and characteristics of VOC emissions from the asphalt pavement, however, several limitations still exist. For example, the database and the calculation methods are insufficient, as the calculation boundary is limited in the production phase and cannot extend to the lifecycle. To help the government to constrain the VOCs emissions within the legal system and draw more attention to the asphalt pavement reduction, this paper first defines the asphalt type, emission type and VOC category, along with the divided phases of the emissions. Then, a general database of VOCs emissions of asphalt pavement is established based on the data from the Environmental Protection Agency (EPA), European Environmental Agency (EEA) and other resources. Thirdly, a calculation method based on the coefficients is proposed to measure and compare the emissions from different types of asphalt under different conditions. Finally, this paper used an example project to test the reliability and feasibility of the method, as well as come up with suggestions for the related stakeholders, such as government, contractors and owners.

Keywords: VOCs · Asphalt · Pavement · Life cycle · Measurement

1 Introduction

Volatile organic compounds (VOCs) are considered to be one of the most significant groups of air pollutants due to their toxic and carcinogenic effects on human health (Lin et al. 2016). In addition, VOCs have a profound influence on the formation of the tropospheric ozone and other oxidants (Yurdakul et al. 2017). Authorities have defined the VOCs from different aspects, e.g. by composition, boiling point and species (United States Environmental Protection Agency 2017; USEPA 2017). The main source of VOCs is natural emission from vegetation such as Isoprene. Anthropogenic sources include fuel production, household products, building materials, furnishings and so on. To control the emissions of VOCs, some countries have created regulations and testing methods for VOC emissions, which have resulted in (1) set emission limits, (2) the

S. El-Badawy and R. Abd El-Hakim (Eds.): GeoMEast 2018, SUCI, pp. 84–93, 2019.
https://doi.org/10.1007/978-3-030-01908-2_7

development of new technology, and (3) estimation, tracking and calculation. In the United States, emissions of VOCs into the atmosphere are regulated by EPA, and each state has its own VOCs emission regulations, caps, and source tracking. In China, the Ministry of Environmental Protection proposed many standards such as integrated emission standards for air pollutants, emission standards of pollutants for the caustic alkali and polyvinyl chloride industries and so on (the Ministry of Ecology and Rnvironment of China 2016). In Europe, the VOC Solvents Emissions Directive is the most influential legislation (Liebscher 2000).

The United States has more than 2.5 million miles of asphalt paved roads and highways. As the most important material, asphalt is a major contributor of VOCs during its production, transportation and paving phases. High temperatures, rainfall and unusual humidity can also lead to the release of VOCs in the maintenance phase. The most commonly used asphalts for pavement construction are grouped into the following general categories: hot mix asphalt (HMA), worm mix asphalt (WMA), cutback asphalt, emulsified asphalt, and so on. On one hand, researchers of VOC emissions from asphalt pavement mainly focused on exploring the impact of catalytic oxidation technologies, usage of additive products, and new asphalt application on the amount and characteristics of VOC emissions (Kamal et al. 2016; Wang et al. 2018; Li et al. 2017). On the other hand, some scholars compared the VOCs emissions between HMA and WMA (del Carmen Rubio et al. 2013), as well as compared and listed the major kinds of VOCs (Iwuoha and Udoh 2016; Lange and Stroup-Gardiner 2005; Myers et al. 2000).

All the available measurement methods above are selective in what they can measure and quantify accurately, as none are capable of measuring all the VOCs that are present. Two kinds of measurement methods are shown in the previous studies to test the total mixed VOCs and classified VOCs. Flame Ionization Detection (FID) and Photo Ionization Detection (PID) methods are used to test the total VOCs (Wilford 2006), while the Gas Chromatograph/Mass Spectrometer (GC/MS) or ambient volatile organic canister sampler (AVOCS) are used to test the classified VOCs (Lange and Stroup-Gardiner 2005). Due to different testing conditions and calculation methods, the results show huge gaps, and cannot widely be promote in the related industries. Therefore, authoritative departments such as the European Environmental Agency (EEA) and the U.S. Environmental Protection Agency (EPA) tried to establish a database and a general coefficient method to estimate the total annual VOC emissions from asphalt plants (USEPA 2000; EEA 2016).

However, several limitations still exist, which are as follows: (1) the database and the calculation method are just for the HMA without considering other types of asphalt, such as foam asphalt and WHA; (2) VOCs are listed as just one of the pollutants, which is not accurately detailed; (3) the boundary is limited to emissions from the production phase, instead of a life-cycle approach. Thus, it is critical to integrate the life-cycle perspective for VOC emissions from asphalt pavement for a comprehensive assessment.

Based on the literature reviews, this paper (1) defines the asphalt type, emission type and VOC category, (2) divides the phases of the emissions, (3) establishes a general database of VOCs emissions of asphalt pavement, (4) proposes the calculation method based the coefficients, (5) compares the emissions from different types of asphalts under different conditions, (6) analyzes the results and comes up with suggestions to the stakeholders, such as government, contractors and owners.

2 Methodology

Based on the EPA data, this paper intends to use the coefficient method to calculate the lifecycle VOCs emissions of asphalt pavement. The required information is mainly from EPA official data, and is supplemented by data from other scholarly' articles. For other unknown factor data, we will take an asphalt plant as an example to conduct actual test data collection. The calculation formulas are largely based on those provided by EPA, and the missing formula will be obtained from the literature review and modified.

Before calculating emissions, the study needs to define the system boundary of objectives, including life-cycle process boundary and emission type boundary. The life cycle process of the VOCs emissions of asphalt pavement is divided into five stages as shown in Fig. 1, namely the (1) asphalt production, (2) asphalt transportation, (3) pavement paving, (4) pavement operation, (5) demolition and reclaim.

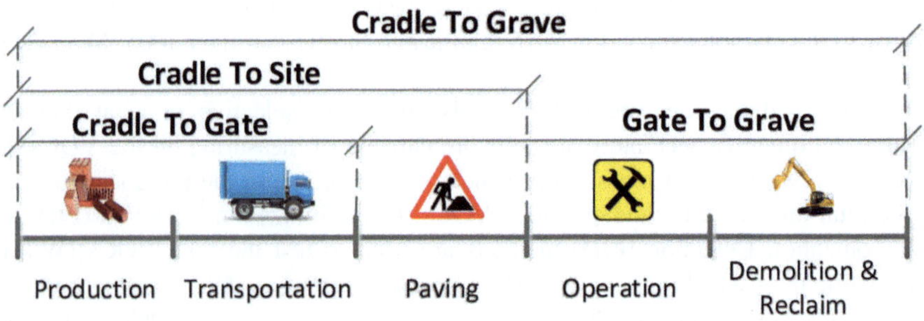

Fig. 1. VOCs emissions of asphalt pavement

Regarding the emission type boundary, there are a number of potential sources of VOCs, including the storage of asphalt and the production of the mix. One source is from the storage of the bulk asphalt in a heated tank. The second source is from stack emissions during production. A third source is from the loading of the mix from the silos into the trucks, and from the trucks into the paver hopper. Since volatilization of odor-related VOCs is an interfacial phenomenon, both the film thickness on the aggregate surface and the surface area of the aggregate should influence VOC concentrations (Lange and Stroup-Gardiner 2005). Therefore, two categories are involved. One is the fugitive emissions of the asphalt itself, which are those not confined to ducts and vents, but emitted directly from the source to the ambient air, such as mixing and storage volatilization, truck loading and transport volatilization, paver loading and paving volatilization, operation volatilization and so on. The other emissions are ducted emissions, which are those vented to the atmosphere through some type of stack, vent, or pipe, including the electricity consumption of the mixing equipment, the fuel consumption of transportation vehicles, and the fuel consumption of the paver and roller and so on.

Regarding the type of asphalts, there are four most widely used types considered in this paper, which are (1) hot mixed asphalt; (2) warm mixed asphalt; (3) foam asphalt; (4) cutback asphalt. Besides these, there is emulsified asphalt among other types. Based on the above definitions and assumptions, the life cycle process of the VOCs emission of asphalt pavement is proposed as follows.

$$E = \sum_{i=1}^{n} (E_{i,f} + E_{i,d}) \tag{1}$$

Where $i = 1, 2, 3, 4, 5$ denotes the life cycle phases including asphalt production, asphalt transportation, pavement paving, pavement operation and maintenance, demolition and reclaim, respectively; f denotes the emission categories of fugitive emissions and d is ducted emission. For example, $E_{4,f}$ represents the fugitive emissions during the operation phase.

(1) Asphalt production phase

Most types of asphalt are produced in asphalt plants, which can be classified by their mixing operation as one of the following: (1) batch mix plants and (2) drum mix plants, as well as parallel-flow drum mix plants, counter-flow drum mix plants and so on. An asphalt plant can be constructed as a permanent plant, a skid-mounted (easily relocated) plant, or a portable plant.

For both batch mix plants and drum mix plants, the estimate includes emissions from the dryer/mixer, load-out operations, asphalt storage, yard (fugitive emissions from loaded trucks), diesel exhaust, paved and unpaved road dust, and aggregate processing (screening, conveyor transfer, and reclaimed asphalt pavement (RAP) crushing). Additionally, for the drum mix plant, the estimate includes emissions from silo filling operations.

According to the above processes, the converted emissions (from lb/yr to g/ton) were estimated using the emission factors developed for the HMA industry in Table 1 based on the assumptions in the EPA AP-42.

Table 1. VOCs emissions for a typical batch/drum mix asphalt plant (Unit: g/ton HMA)

Batch mix plant	Amount	Drum mix plant	Amount
Dryer, hot screens, and mixer	3.72	Dryer	14.51
Load-out operations	1.77	Load-out operations	1.77
–	–	Silo filling operations	5.44
Storage tank	0.15	Storage tank	0.15
Yard	0.50	Yard	0.50
Total	6.80	Total	22.68

(Source: EPA AP-42)

Therefore, the VOCs emissions of production phase are as follows:

$$E_1 = M_a \times F_a \quad (2)$$

Where E_1 is the VOCs emissions during production phase; M_a is the amount of asphalt produced (Unit: ton); F_a denotes the VOCs emission factor (Unit: g/ton), if the batch mix plant is chosen, the factor value is 6.8 g/ton; if the drum mix plant is chosen, the factor value is 22.68 g/ton.

For the production of foam asphalt, the general method is almost the same as HMA, however, the steam device and additives are added in the mix stage. In addition, the aggregate grading and tapping temperature are different from that of HMA. More on-site testing data is needed. The emission factor of load-out operations and silo filling of a drum mix or batch mix plant can be calculated as follows.

$$F_y = \alpha(-V)(w)e^{((0.0251)\ (T\ +\ 460) - 20.43)} \quad (3)$$

Where α is the initial coefficient of TOC, the default values are 0.0172 and 0.0504 for load-out operations and silo filling phase; T is the temperature, default value for HMA is 325°F, 240°F for the foam asphalt, and 250 for WMA (Rubio et al. 2012); V is the percentage of asphalt loss-on-heating, default value is −0.5; w is percentage of VOC in the TOC. The default values are 94% for the load-out and yard, and 100% for the silo filling and asphalt storage tank emissions.

The yard emissions, namely the emission after load-out phase, usually last for 8 min. The average emission rate was 0.19 lb/hr of TOC. The TOC emissions for the 8-minute period immediately following load-out (yard emissions) can be estimated using an emission factor of 0.00055 kg/Mg (0.0011 lb/ton) of asphalt loaded.

(2) Transportation phase

In this phase, due to the short time of exposure to the open air, the volatilization of the VOCs emission is negligible. The ducted emissions are mostly from the transportation vehicle. The EMEP/EEA air pollutant emission inventory guidebook 2016 gives very detailed data with units of both gram VOCs per kilometer and per kilogram of fuel usage (Ntziachristos and Samaras 2017). In this paper, we chose the former as the standard for the transportation phase Therefore, the VOCs emission during the transportation phase is as follows:

$$E_2 = L_p \times F_p \quad (4)$$

Where E_2 is the VOCs emissions during transportation phase; L_p denotes the distance between the asphalt plant and the project site (Unit: km), F_p is the emission factor of VOC by different type of transportation vehicle (Unit: g/ton/km); p is the type of vehicle and fuel. The VOC factors of light commercial and heavy-duty vehicles are shown in Tables 2 and 3.

GREETTM proposed that the lifetime mileage-weighted average air pollutant emission factors for gasoline and diesel single-unit short-haul trucks for year 2018 are

Table 2. NMVOC of light commercial vehicles (unit: g/km)

Petrol	Conventional	3.44
	Euro 1 - 93/59/EEC	0.614
	Euro 2 - 96/69/EEC	0.304
	Euro 3 - 98/69/EC I	0.189
	Euro 4 - 98/69/EC II	0.128
	Euro 5 – EC 715/2007	0.096
	Euro 6 up to 2017	0.096
	Euro 6 2018–2020	0.096
	Euro 6 2021+	0.096
Diesel	Conventional	0.133
	Euro 1 - 93/59/EEC	0.141
	Euro 2 - 96/69/EEC	0.149
	Euro 3 - 98/69/EC I	0.094
	Euro 4 - 98/69/EC II	0.035
	Euro 5 – EC 715/2007	0.035
	Euro 6 up to 2017	0.035
	Euro 6 2018–2020	0.035
	Euro 6 2021+	0.035

Note: Given as THC-CH4

0.9124 and 0.0772 g/mile. (Source: Updated Emission Factors of Air Pollutants from Vehicle Operations in Using MOVES)

(3) Paving phase

In this phase, most VOCs emissions are from the paving machine and equipment, including pavers, compactors and so on. Based on the EMEP/EEA air pollutant emission inventory guidebook for 2016 (Ntziachristos and Samaras 2017), there are some certain relations between VOC emissions and fuel properties for vehicles. The density of fuel, Sulphur content, polycyclic aromatics content, cetane number and back end distillation are considered as the key factors. For example, the relation between emission and fuel properties of diesel light commercial is as follows.

$$\begin{aligned} \text{Fcorr} = {} & -0.293192 + 0.0006759 * \text{DEN} - 0.0007306 * \text{PAH} - 0.0032733 * \text{CN} \\ & - 0.000038 * \text{T95} \end{aligned}$$

Where DEN = Density at 15 °C [kg/m^3], S = Sulphur content in ppm, PAH = Polycyclic aromatics content in %, CN = Cetane number, T95 = Back end distillation in °C.

Therefore, the VOCs emission during transportation phase is as follows:

$$E_3 = \sum_i \left(\sum_m (FC_{i,m} \times EF_{i,m})\right) \quad (5)$$

Where E_3 = emission of VOC [g], $FC_{i,m}$ = fuel consumption of vehicle category i using fuel m [kg], and $EF_{i,m}$ = VOC emission factor for vehicle category i and fuel m [g/kg]. So far, the relationship between the emissions and the fuel consumption of pavers, compactors and so on remain unknown. More on-site data are needed.

(4) Operation phase

Table 3. NMVOC of heavy-duty vehicles (unit: g/km)

Petrol > 3.5 t	Conventional	5.25
Diesel <= 7.5 t	Conventional	1.07
	Euro 1 - 91/542/EEC I	0.193
	Euro 2 - 91/542/EEC II	0.123
	Euro 3 - 2000	0.115
	Euro 4 - 2005	0.005
	Euro 5 - 2008	0.005
	Euro 6	0.005
Diesel 7.5 - 16 t	Conventional	0.776
	Euro 1 - 91/542/EEC I	0.326
	Euro 2 - 91/542/EEC II	0.207
	Euro 3 - 2000	0.189
	Euro 4 - 2005	0.008
	Euro 5 - 2008	0.008
	Euro 6	0.008
Diesel 16 - 32 t	Conventional	0.486
	Euro 1 - 91/542/EEC I	0.449
	Euro 2 - 91/542/EEC II	0.29
	Euro 3 - 2000	0.278
	Euro 4 - 2005	0.01
	Euro 5 - 2008	0.01
	Euro 6	0.01
Diesel > 32 t	Conventional	0.534
	Euro 1 - 91/542/EEC I	0.51
	Euro 2 - 91/542/EEC II	0.326
	Euro 3 - 2000	0.308
	Euro 4 - 2005	0.012
	Euro 5 - 2008	0.012
	Euro 6	0.012

Note: Given as THC-CH4

In this phase, the major source of VOC is from cutback asphalt, while only minor amounts of VOCs are emitted from emulsified asphalts HMA and other similar types of asphalt cement. Some studies also proposed that the VOC concentrations were extremely low, when measured in both the laboratory and on the construction sites (Lin et al. 2016). Therefore, only VOCs emission volatized from cutback asphalt is calculated. The VOC emissions depend on the amount of VOCs in the diluent used in asphalt. Thus, the amount of each asphalt type and the diluent content must also be determined. Cutback asphalts fall into three broad categories: rapid cure (RC), medium cure (MC), and slow cure (SC) road oils.

For any given amount of cutback asphalt, total emissions are believed to be the same, regardless of stockpiling, mixing, and application times. The two major variables affecting both the quantity of VOCs emitted and the time over which emissions occur are the type and the quantity of petroleum distillate used as a diluent. The formula of emissions from diluent of cutback asphalt is as follows:

$$E_{4,f} = M_d \times k \tag{6}$$

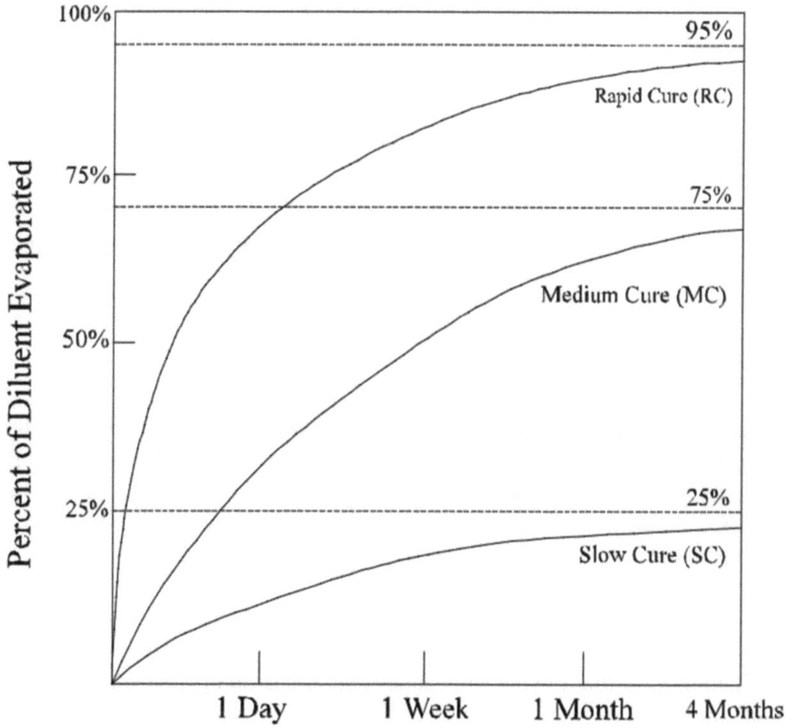

Fig. 2. Short-term VOCs emission curve (source: USEPA AP-42)

Where $E_{4,f}$ denotes the VOC emissions of fugitive emissions during operation phase; M_d represents the amount of diluent (Unit: liter); k is the percentage of the diluent evaporated. It can be estimated as 95% of diluent weight for RC cutback asphalts, 70% for MC, and about 25% for SC. The typical diluent density is assumed to be 0.7, 0.8 and 0.9 kg/l of rapid, medium and slow cure asphalts, respectively. Of course, if actual density values are known from local records, they should be used in the above equations rather than the estimated values.

Note that evaporation to these levels occurs over a period of about four months. Evaporation curves are provided in Fig. 2. About 75% by weight of diluent evaporates in the first day following application of rapid cure cutback asphalts whereas it takes about one week for 50% by weight of the diluent to evaporate from medium cure cutback asphalts.

(5) Demolition and reclaim phase

There are five methods of recycling including cold planning, hot recycling, hot in-place recycling, cold in-place recycling, and full depth reclamation. All except hot recycling occur at the location where the paving is taking place, although material removed during cold planning may be processed at an asphalt plant. In this phase, most VOCs emissions are from the paving machine and equipment, including crushers, trucks and so on. Therefore, the VOCs emission during this phase is calculated in almost the same way as that during the paving phase:

$$E_5 = \sum_i \left(\sum_m (FC_{i,m} \times EF_{i,m}) \right) \tag{7}$$

Where E_5 = emission of VOC [g], $FC_{i,m}$ = fuel consumption of vehicle category i using fuel m [kg], $EF_{i,m}$ = VOC emission factor for vehicle category i and fuel m [g/kg]. So far, the relationship between the emission and fuel consumption of crushers remain unknown. More on-site data are needed.

3 Conclusions

In summary, the contributions of this paper include: (1) defining and dividing the phases of lifecycle VOCs emission of asphalt pavement; (2) summarizing the emission characteristics of four typical asphalts; (3) establishing the calculation method for lifecycle VOCs emission of asphalt pavement based on the coefficient method.

However, some limitations can be seen in the study, such as: (1) lack of the emission factor of WMA and foam asphalt during the production phase; (2) lack of emission factor of machines and equipment during paving and demolition phase; (3) lack of the latest data or case study for the whole life cycle of an asphalt pavement project. Therefore, in future work, this paper may focus on (1) collecting the emission factor data by testing on-site; (2) taking a pavement project as an example to verify the feasibility of the calculation method; (3) proposing a trading system of VOCs in the construction industry.

References

EEA: Air pollutant emission inventory guidebook- Emission factors (2016)

del Carmen Rubio, M., Moreno, F., Martínez-Echevarría, M.J., Martínez, G., Vázquez, J.M.: Comparative analysis of emissions from the manufacture and use of hot and half-warm mix asphalt. J. Clean. Prod. **41**, 1–6 (2013)

Iwuoha, G.N., Udoh, C.: Volatile Organic Compounds and Heavy Metals in Asphalt Concrete Used in Road Surfacing of East West Road, Nigeria. J. Chem. Soc. Nigeria **41**(1) (2016)

Kamal, M.S., Razzak, S.A., Hossain, M.M.: Catalytic oxidation of volatile organic compounds (VOCs)–a review. Atmos. Environ. **140**, 117–134 (2016)

Lange, C.R., Stroup-Gardiner, M.: Quantification of potentially odorous volatile organic compounds from asphalt binders using head-space gas chromatography. J. Test. Eval. **33**(2), 101–109 (2005)

Li, L., Wu, S., Liu, G., Cao, T., Amirkhanian, S.: Effect of organo-montmorillonite nanoclay on VOCs inhibition of bitumen. Constr. Build. Mater. **146**, 429–435 (2017)

Liebscher, H.: Economic solutions for compliance to the new European VOC Directive. Prog. Org. Coat. **40**(1–4), 75–83 (2000)

Lin, S., Hung, W., Leng, Z.: Air pollutant emissions and acoustic performance of hot mix asphalts. Constr. Build. Mater. **129**, 1–10 (2016). https://doi.org/10.1016/j.conbuildmat.2016.11.013

Myers, R., Shrager, B., Brooks, G.: Hot Mix Asphalt Plants Emission Assessment Report (2000)

Ntziachristos, L., Samaras, Z.: The EMEP/EEA air pollutant emission inventory guidebook 2016. European Environment Agency (2017)

Wilford, P.: VOC Detection and Measurement Techniques (2006)

Rubio, M.C., Martínez, G., Baena, L., Moreno, F.: Warm mix asphalt: an overview. J. Clean. Prod. **24**, 76–84 (2012)

The Ministry of Ecology and Environment of China.: Emission standard of pollutants for the caustic alkali and polyvinyl chloride industries GB15581-2016 (2016)

USEPA: Air Emissions Inventories. United States Environmental Protection Agency (2000)

USEPA: AP-42: Compilation of Air Emissions Factors, United States Environmental Protection Agency (2009)

USEPA: Technical Overview of Volatile Organic Compounds. United States Environmental Protection Agency (2017)

Wang, T., Xiao, F., Zhu, X., Huang, B., Wang, J., Amirkhanian, S.: Energy consumption and environmental impact of rubberized asphalt pavement. J. Clean. Prod. **180**, 139–158 (2018)

Yurdakul, S., Civan, M., Kuntasal, Ö., Doğan, G., Pekey, H., Tuncel, G.: Temporal variations of VOC concentrations in Bursa atmosphere. Atmos. Pollut. Res. (2017)

Foamed Bitumen Base for Expedient Airport Pavement Upgrade: A Case Study on Whitsunday Coast Airport

Greg White(✉)

Airport Pavement Research Program, University of the Sunshine Coast, Sippy Downs, QLD, Australia
gwhite2@usc.edu.au

Abstract. Foamed bitumen base is one method for producing a bitumen stabilised granular material for pavement construction. Foamed bitumen relies on the rapid expansion (foaming) of hot bitumen when introduced to cold water, temporarily reducing the viscosity of the bitumen, allowing mixing, placement and compaction. The resulting material is fast to construct, stiffer than crushed rock, moisture resistant and almost immediately able to support traffic, making it valuable for expedient pavement expansion and strengthening works. Airports with single runways often require expedient pavement construction methods and this paper presents foamed bitumen base through a case study on the runway strengthening at Whitsunday Coast Airport in Australia. Foamed bitumen base was produced ex-situ, initially using new crushed rock, prior to transitioning to the existing pavement materials. Challenges included the inability to recover existing pavement material in a representatively pulverised condition during the mixture design phase and the high sensitivity of the foamed bitumen base to production moisture content. Isolated underlying unsuitable materials were also encountered and this impacted the overall productivity and program. However, the relative density and resilient modulus of all samples exceeded the relevant requirements and foamed bitumen base is recommended to other airports requiring expedient pavement upgrade solutions.

1 Introduction

Many regional or remote runways were designed for aircraft much smaller than the largest aircraft operating at the time. As passenger movements increase, the smaller aircraft are replaced by larger aircraft, which overload the runway. As a result, structural upgrade or strengthening is often required. Further, newer aircraft are often wider and longer than previous aircraft, requiring new pavement to be constructed immediately adjacent existing runways (Forsyth 2005). Many airports are 24 hour per day operations, meaning extended runway closures to facilitate upgrade works are not usually practicable. This is especially the case at single-runway airports. However, even at multi-runway airports, the closure of one runway often has significant impact on the overall efficiency of airport operations. It follows that runway upgrade work is often performed at night, with the runway returned to service each morning. The period allowed to perform such upgrade work is ever reducing as airports become busier and

S. El-Badawy and R. Abd El-Hakim (Eds.): GeoMEast 2018, SUCI, pp. 94–111, 2019.
https://doi.org/10.1007/978-3-030-01908-2_8

this combination of circumstances requires expedient construction materials and methods for many runway upgrades.

Foamed Bitumen Base (FBB) is an expedient pavement construction and rehabilitation technology that has gained significant interest for airport pavement upgrades in Australia in recent years (White 2018b). FBB was produced from both a new crushed rock and from an existing pavement material as part of the reconstruction of the Whitsunday Coast Airport (WCA) upgrade in 2017. This paper presents the use of FBB for expedient aircraft pavement strengthening, using the data collected from the WCA project as a case study. Mixture design and production consistence are considered, as well as constructability issues.

2 Background

2.1 Foamed Bitumen Base

FBB is one of three methods for producing what is effectively a bitumen stabilised granular material. FBB relies on foaming of bitumen (like dropping water into hot oil) to temporarily reduce the viscosity of the bitumen, allowing mixing, placement and compaction of the stabilised material. Bitumen emulsion stabilised base (ESB) is similar, but takes advantage of the low viscosity of the emulsified bitumen to achieve a similarly workable product. Bitumen treatment base (BTB) is also similar, but relies on heat to reduce the bitumen viscosity. BTB is effectively a low bitumen content asphalt produced from unfractionated base or sub-base quality granular material and is produced in a hot-mixed asphalt plant.

FBB offers some advantages compared to BTB and ESB. ESB is slow curing and while the long-term properties of ESB and FBB are expected to be similar, the short-term trafficability of ESB is unlikely to be suitable for expedient pavement upgrades required for airports. BTB is faster curing, due to the stiffening of the bituminous binder associated with the reduction in temperature in the hours following production. However, BTB relies on the availability of a hot-mixed asphalt production plant, which is costly to mobilise to remote areas when not otherwise required. Furthermore, BTB is not able to be produced in-situ, reducing its flexibility for re-use of existing pavement materials.

FBB is often compared to cement stabilised base course (CTB). However, compared to CTB, the risk of FBB cracking, which will reflect through the bituminous surfacing, is almost completely mitigated. This is especially important for pavements with either a thin asphalt surfacing or a sprayed seal surfacing. FBB also allows greater layer thicknesses and multiple layers of FBB adhere to each other more reliably that CTB layers.

FBB is now recognised as a cost-effective expedient pavement construction technology (White 2014). The properties of FBB that make it suitable for expedient pavement construction and rehabilitation include (White 2018b):

- Significant shelf-life after production.
- Rapid construction in multiple layers.
- High stiffness providing an efficient contribution to structural capacity.

- Early trafficability due to ambient production temperature.
- Moisture resistance resulting from the bound nature of the material.

FBB can be produced from new crushed rock or from existing pavement materials, whether a crushed rock, a crushed natural gravel or an uncrushed gravel. It can be produced in-situ, using a specialised pulveriser/stabiliser, or ex-situ, in a pugmil. From a structural perspective, FBB is similar to asphalt, except that it is produced at ambient temperatures, requires significantly less bituminous binder and does not require granular material fractionation.

2.2 Whitsunday Coast Airport Upgrade

WCA is located approximately 15 km south of the town of Proserpine in the Whitsunday Coast region of tropical north Queensland. It is the primary airport for the Airlie Beach area, which is a major Queensland tourist destination as the main gateway to the Whitsunday Islands and the Great Barrier Reef. The region is also a significant sugar cane growing and processing centre. The airport supports both the sugar cane and tourism industries and forms part of the region's critical infrastructure.

The airport is lit and capable of accommodating 24-hour flight operations with one main runway (2,075 m by 45 m), a parking apron capable of accommodating up to four commercial aircraft simultaneously, as well as a connecting taxiway. Other infrastructure includes general aviation areas and a passenger terminal building.

In 2012 the runway at WCA was in poor condition, exhibiting significant roughness and cracking as a result of continuous overloading and inundation associated with flood events in 2010 and 2012. A significant repair was affected in late 2012 and a major pavement rehabilitation was planned for the future. That major rehabilitation was complete over three months in mid 2017 and is the subject of this case study.

Operations at the airport required the upgrade to be affected between 1800 and 0600 nightly, with the pavements reinstated to a serviceable condition each morning. In response to a desire to recycle and reuse materials wherever possible, as well as providing a cost-effective upgrade solution and to increase the strength and moisture resistance of the pavement, the following design was adopted:

- Stabilisation of the upper 250 mm of the existing pavement, comprising a 50 mm thick asphalt surface on 300 mm of natural gravel, with foamed bitumen to create FBB.
- 100 mm asphalt surfacing.
- Grooving, linemarking, airfield lighting modifications, flank regrading and other ancillary works.

Due to the delay associated with the processing and testing of excavated existing pavement materials and subsequent production of the FBB, the first portion of the work was performed with a new quarried crushed rock. As discussed later, verification of the processing and FBB production with the existing pavement materials was challenging. Consequently, the change from new crushed rock to the existing pavement material for FBB production was delayed by approximately one week.

It is worth noting that the works were delivered via a design and construct procurement model (White 2018a) and both the FBB and the asphalt specifications were performance-based (White 2017). This allowed the experience and expertise of the contractor to be incorporated during the finalisation of the design solution, as well as the mixture design and production of the FBB.

3 In-Situ Versus Ex-Situ FBB Production

It was initially expected that the foamed bitumen stabilisation of the existing pavement would be performed in-situ using a specialised pulveriser/stabiliser. This reflected in-situ processing being most common and highly cost-effective whenever an existing base course material is stabilised (White 2018b). However, a number of risks were identified:

- Inability to sort and test the existing natural gravel base course.
- Reduced ability to identify and treat unsuitable underlying pavement material and subgrade.
- Potential for a mechanical failure preventing completion of the required scope during the shift.
- Inability to produce all the FBB required for the shift prior to commencement of the shift and the associated risk of not returning the pavement to service by the required time.
- Likelihood of excessive moisture content at the commencement and completion of each in-situ stabilisation run.
- Reduction in ability to treat or supplement the existing pavement material without a significant reduction in the nightly productivity.

Consequently, ex-situ production in a specialised pugmil, located in the on-site work compound, was adopted. The ex-situ option was associated with a significant cost premium. However, the reduced technical risk and increased confidence that the runway would be returned to a serviceable condition by the end of each work period overrode the financial impost.

4 Mixture Design

Two FBB mixture designs were performed, one for the new crushed rock and one for the existing pavement material. Due to the complexity associated with recovering samples of the existing pavement consistent with the material excavated during the works, the existing pavement material FBB mixture design presented a challenge.

First, samples of the new crushed rock were obtained from the quarry, characterised and trial FBB mixtures were produced in a small-scale laboratory bitumen foaming and mixing device. The trial mixtures were intended to optimise the bitumen and lime contents with the overriding aims of the mixture design process being to (Austroads 2011):

- Provide an appropriate aggregate grading, with adequate fine (less than 0.075 mm in diameter) material to combine with the foamed bitumen.
- Achieve a soaked (by 10 min of vacuum saturation) resilient modulus (measured by indirect tension at 20 °C) of not less than 1,500 MPa.
- Produce appropriate resistance to moisture, with a reduction in modulus after soaking of not more than 40%.

In all cases the pre-soaked (cured) modulus was measured after three days of accelerated curing at 60 °C. It is worth noting that the structural design of the pavement upgrade was based on a conservative FBB modulus of 800 MPa. This allowed for variability in FBB production during the work, as well as a reduction in modulus during hot weather, associated with the change in bitumen viscosity.

Three new crushed rock samples for FBB mixture design exhibited a consistent grading that was well inside the limits recommended by Austroads (2011) as shown in Fig. 1. Three mixtures were prepared in the laboratory with different bitumen contents (Table 1). The existing pavement material FBB was intended to replicate the properties of the new crushed rock design. Small samples of the existing asphalt and the underlying existing natural gravel were recovered from the pavement by coring through the asphalt and auguring through the granular base course. In the laboratory, the samples were pulverised and combined in a ratio of 4:1 (gravel to asphalt) to replicate the composition of the depth of existing pavement to be stabilised, which included 50 mm of asphalt surface and 200 mm of natural gravel. Based on experience, it was expected that the natural gravel would require the same or more bitumen than the new crushed rock. Consequently, only two existing material FBB trial mixtures were prepared and tested (Table 2). The grading of the combined existing pavement material was more variable than the new crushed rock and one sample fell outside the recommended grading limits (Fig. 2). However, the portion of fine material, which is recognised as being critical to FBB performance, was within the recommended range.

For the FBB produced with the new crushed rock, the cured modulus did not change significantly with increasing bitumen content, although the soaked modulus increased by approximately 50% with an increase in bitumen content from 2% to 3% (Fig. 3). However, the additional increase in bitumen content to 4% did not increase the soaked modulus any further. Consequently, 3% bitumen was adopted for the new crushed rock FBB design. Both the cured and soaked existing pavement FBB modulus were insensitive to the bitumen content over the range tested (Fig. 3). Furthermore, the existing pavement FBB moduli values at 3% bitumen content were similar to the new crushed rock FBB values. Therefore, 3% bitumen content was also adopted for the FBB containing the existing pavement material.

Figure 3 indicates that there remained opportunity to further optimise the design of the FBB mixtures. Bitumen contents of 2–2.5% were worth assessment for the existing pavement material while 2.5% bitumen content was of interest for the new crushed rock FBB. However, the limited ability to recover large volumes of material from the existing pavement and the overall project milestones prevented further FBB optimisation in the laboratory.

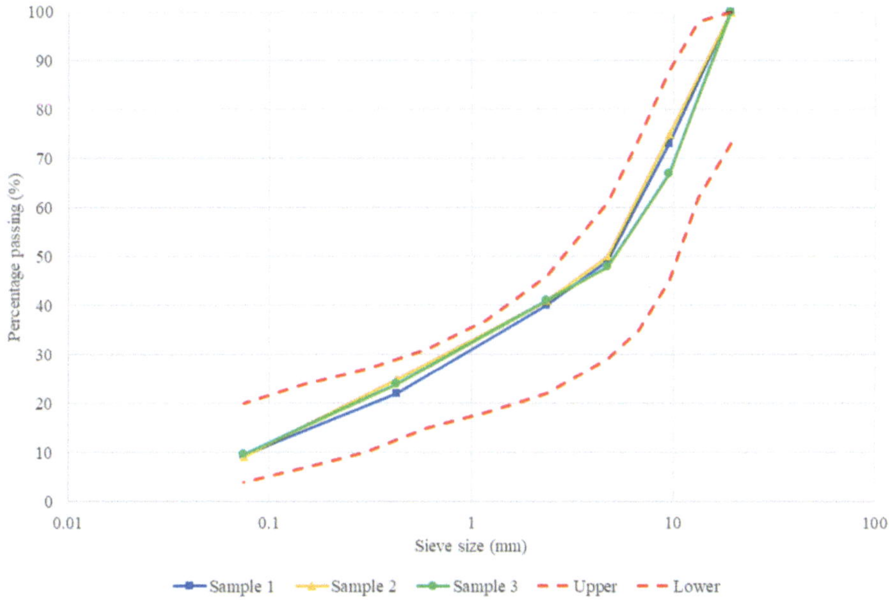

Fig. 1. New crushed rock aggregate grading

Table 1. Summary of new crushed rock FBB designs

Bitumen content	Lime content	Cured modulus	Soaked modulus	Modulus reduction
2%	1%	2,382 MPa	985 MPa	59%
3%	1%	2,358 MPa	1,451 MPa	38%
4%	1%	2,469 MPa	1,505 MPa	39%

All FBB modulus testing was performed in indirect tension at 20 °C.

Table 2. Summary of existing material FBB designs

Bitumen content	Lime content	Cured modulus	Soaked modulus	Modulus reduction
3%	1%	2,289 MPa	1,631 MPa	26%
4%	1%	2,241 MPa	1,652 MPa	26%

5 Production

Not dissimilar to other pavement materials that are produced and constructed on site, production of material that is consistent with the laboratory mixture design is important for pavement performance and durability. For FBB, the critical production factors are:

- Consistent unstabilised material properties.
- Consistent FBB composition.

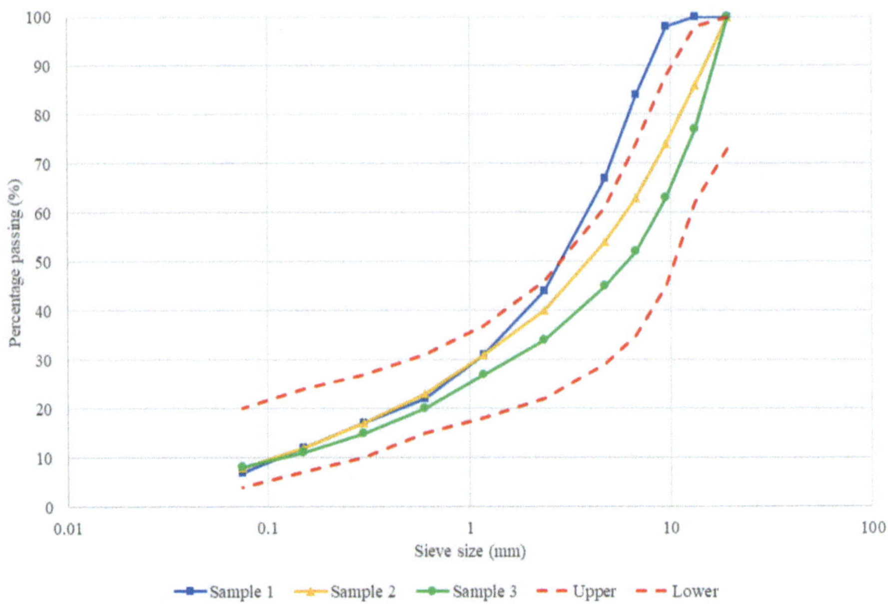

Fig. 2. Existing pavement aggregate grading

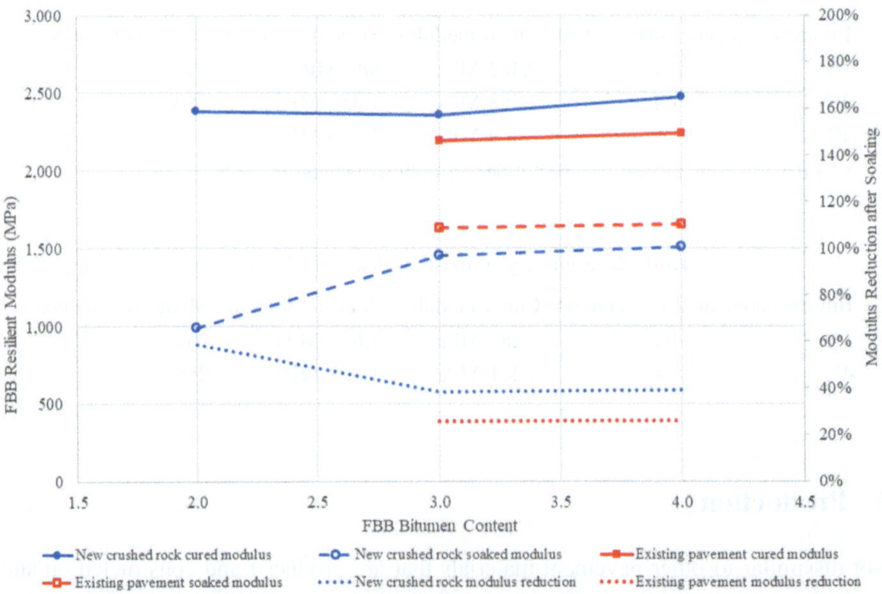

Fig. 3. Affect of bitumen content on FBB modulus

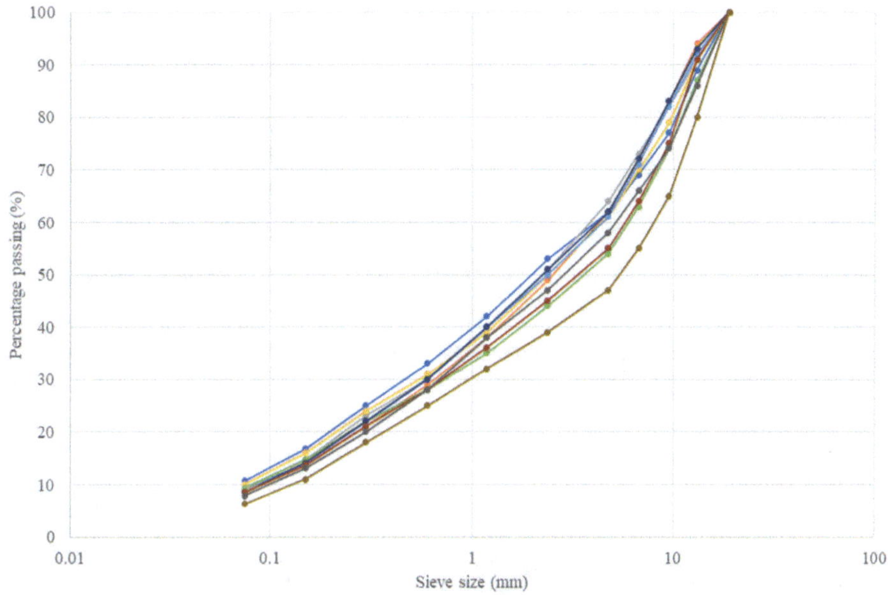

Fig. 4. New crushed rock FBB aggregate grading during production

- Consistent FBB properties.

Because the new crushed rock was a controlled quarry product, the produced material was consistent with the samples used for mixture design and remained consistent throughout the project. The FBB production composition was also consistent, and as a result the critical FBB grading was consistent (Fig. 4).

The existing material was more challenging. Unlike the material recovered for the laboratory mixture design, the existing pavement was excavated by cold planing machines. Due to the expedient nature of the works, the existing pavement was excavated with four profilers operating a maximum capacity and speed. Consequently, the asphalt surface was recovered in large pieces and was not thoroughly pulverised. To produce a material that was suitable for FBB production:

- Material exceeding 200 mm was screened off and disposed of, because it did not fit into the RAP crusher.
- Material exceeding 20 mm was subsequently screened off (Fig. 5) and crushed.
- The crushed RAP was reincorporated into the natural gravel, which already contained RAP smaller than 20 mm that had been pulverised by the cold planers during the excavation process (Fig. 6).
- The recombined material was tested prior to FBB production.

Fig. 5. Large RAP particles for crushing and reincorporation

This process created two issues:

- The RAP particles exceeding 200 mm represented around 20% of the existing pavement material, which required replacement to ensure the required volume of FBB was available.
- The crushed RAP, when reintroduced to the natural gravel, effectively bound all of the fine material, pushing the 'effective' or 'apparent' grading well outside of the recommended limits (Fig. 7).

The lack of fine material in the processed existing pavement material resulted in a deficiency of fines to take-up the foamed bitumen during FBB production and the resulting FBB was unstable and was not readily compacted. To overcome these issues, a sandy loam (Fig. 8) was introduced at a rate of 20% of the existing pavement material (Fig. 9). This corrected the apparent grading of the processed existing pavement material (Fig. 10) as well as replacing the lost oversized RAP material. The addition of the sandy loam produced a FBB that was readily compacted and stable. However, the grading tests continued to artificially appear deficient in fine material because it remained bound by the bituminous mastic in the pulverised RAP (Fig. 11). The primary other FBB production property of interest was the production moisture content, which averaged 3.9% and was adequately consistent throughout the project, with a standard deviation of 0.4%, giving a coefficient of variation of 11% over the 38 samples tested.

Fig. 6. Existing pavement material

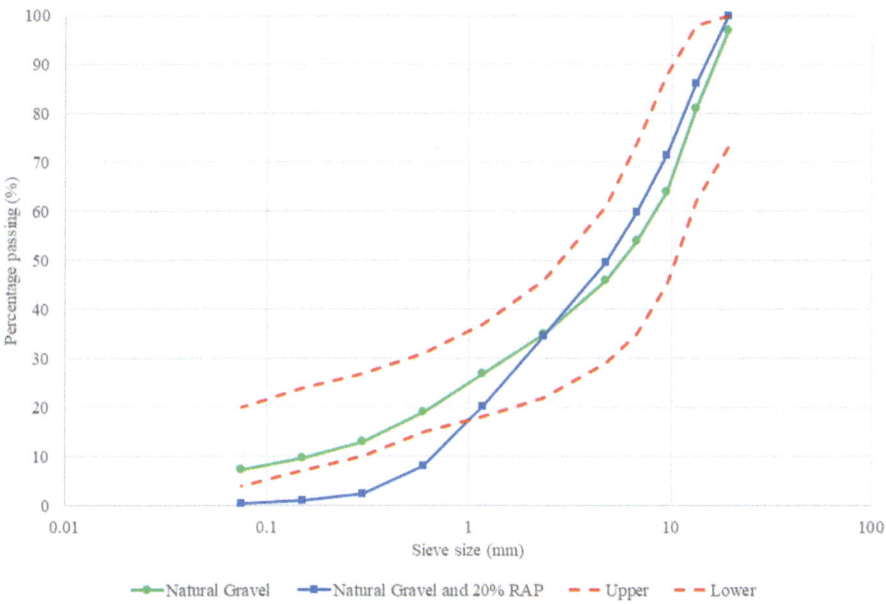

Fig. 7. Apparent grading of existing material with and without RAP

Fig. 8. Sandy loam

Arguably, the modulus is the most important FBB property because it is fundamental to the structural design of the pavement. Samples were obtained, cured for three days and tested for resilient modulus (at 20 °C) before and after vacuum soaking. This protocol was consistent with the mixture design process (Austroads 2011). A total of 108 samples were tested and all exceeded the 800 MPa modulus assumed in the structural design, by at least 100 MPa. On average the FBB produced from the new crushed rock had a higher modulus than the samples produced from the existing pavement materials, but the existing pavement material FBB had a less variable modulus (Table 3). On average, the reduction in modulus due to soaking was 21% and 23% for the new crushed rock and existing material samples, respectively. The reduction in modulus on soaking was similar to that observed during the mixture design process (Tables 1 and 2). Furthermore, the results for the new material and existing material were comparable, with the new material having approximately 40% higher cured and soaked modulus than the existing material. This likely reflects the better inter-particle friction associated with the new material, which was fully crushed. In contrast, the variability of both the cured and soaked soaked modulus results was lower for the existing material than for the new material.

Fig. 9. Existing pavement material after additional of sandy loam

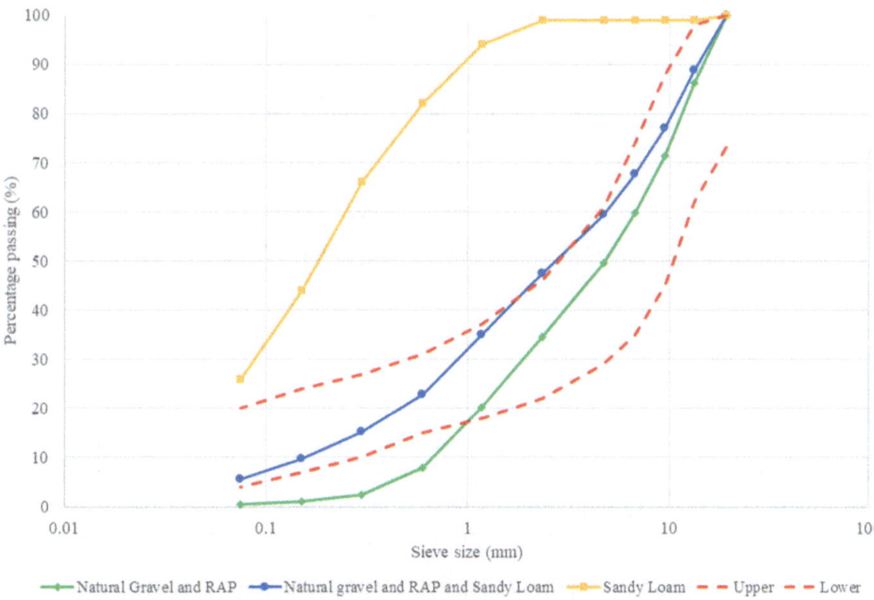

Fig. 10. Correction of apparent grading by addition of sandy loam

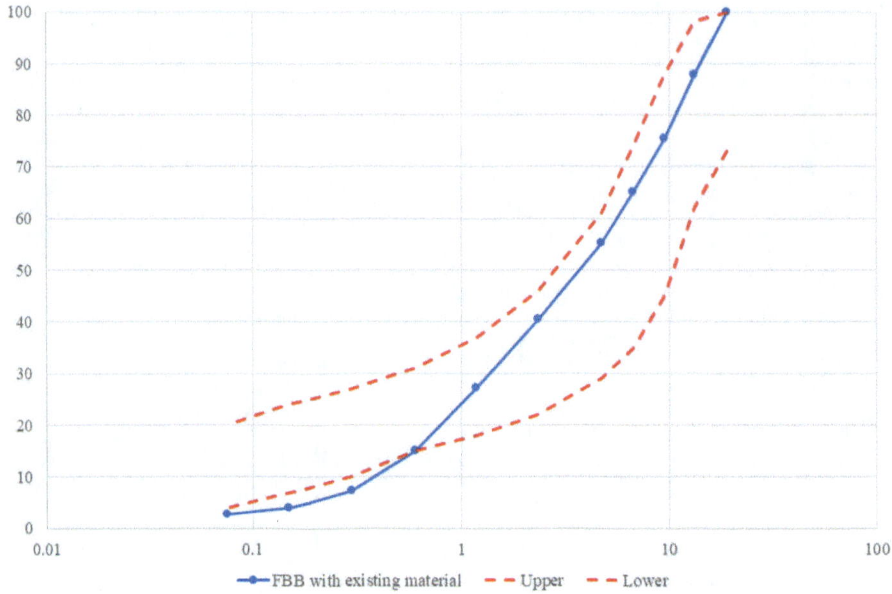

Fig. 11. Average apparent FBB production grading with existing material

Table 3. Summary of produced FBB modulus

Statistic	New crushed rock		Existing material	
	Cured	Soaked	Cured	Soaked
Minimum	1,013	911	1,183	938
Average	2,898	2,285	2,066	1,586
Maximum	4,459	3,738	3,767	2,813
Std. Dev.	986	884	571	454
CoV	34%	39%	28%	29%

Std. Dev. is the standard deviation and CoV is the coefficient of variation.

6 Construction

During construction, both FBB mixtures were paved in a single 250 mm thick layer using two heavy duty gravel pavers and a combination of steel drum and pneumatic tyred rollers. The primary requirements relating to construction were the density and thickness of the FBB. Density is important for avoiding post-construction densification and was specified as not less than 98% of the density achieved using modified Proctor compaction in the laboratory. Layer thickness is critical for the structural capacity of

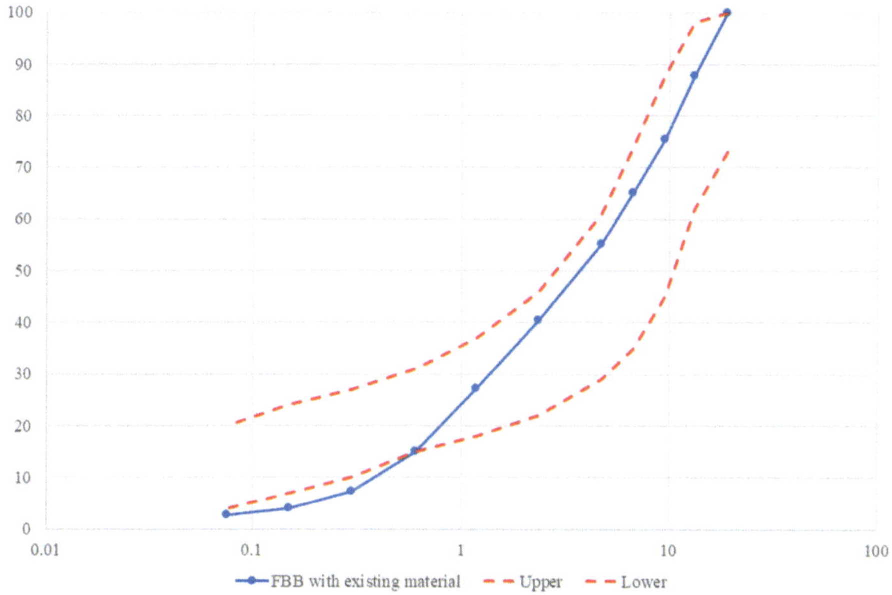

Fig. 12. Constructed FBB relative density

the pavement and a construction tolerance of 20 mm was expected around the targeted nominal thickness of 250 mm.

For the FBB constructed with the new crushed rock, the achieved relative density ranged from 96–102% with an average of 99%, while the relative density of the existing material FBB ranged from 98–106% with an average of 102% (Fig. 12). These statistics were calculated from 54 and 82 test results for the new crushed rock and existing material FBB mixtures, respectively. It is likely that the FBB produced from the existing material was more readily compacted due to the addition of the fine sandy loam material, which comprised naturally weathered and rounded particles instead of angular crushed rock fines.

The thickness of the FBB was more complex and did not vary greatly between the FBB produced with the new crushed rock and that produced with the existing pavement material. However, the FBB thickness was expected to be consistently around 250 mm because of the nominal 250 mm deep excavation that was subsequently filled back to the existing surface level with FBB. This was not the case with the FBB thickness varying from 199 mm up to 285 m, but averaging 231 mm (Table 4). This reflected the construction staging, which included two shifts in the central 27 m followed by a shift in the outer 18 m (9 m each side of the central area) to maximise productivity, extend the distance between transverse joints and maintain an access route for haulage trucks around the work zone. Because the FBB required a thin asphalt surface to be constructed each night until all the FBB was complete and the new surface was constructed over the top, the FBB was thinner at the edges and the lost FBB thickness was replaced with additional asphalt thickness. Consequently, the thin asphalt surfacing ranged from 20 mm to 73 mm and averaged 47 mm, reflecting the

Table 4. Summary of FBB and thin asphalt layer thicknesses

Statistic	FBB	Thin Asphalt	Combined
Minimum (mm)	199	20	235
Average	231	47	283
Maximum (mm)	285	73	324
Std. Dev. (mm)	15	14	21
CoV	7%	29%	8%
Number of results	39	39	39
TARGET (mm)	250	30	270

nominal 30 mm thickness in addition to 14 mm to replace the reduced FBB thickness. Overall, the combined FBB and thin asphalt surface thickness averaged 283 mm compared to the 280 mm target. When the permanent wearing course was added, the combined pavement thickness was just 1 mm less than the designed 350 mm total pavement thickness. However, significant variance remained between the intended and measured FBB/asphalt thicknesses and it is likely that this reflected some movement in the underlying subgrade during FBB compaction.

During construction the contractor reported occasional difficulty achieving FBB compaction due to excess moisture content. Target moisture contents for FBB are lower than for crushed rock due to the lubricating effect of the bituminous material, with 60–80% of the optimum moisture content of the granular material recommended

Table 5. Summary of FBB moisture content percentage of optimum

Statistic	New crushed rock	Existing material
Minimum	41%	27%
Average	73%	61%
Maximum	102%	113%
Std. Dev.	14%	17%
CoV	19%	28%
Number of results	54	82

(Austroads 2011). Moisture contents varied for both FBB mixtures (Table 5) but the moisture content was not correlated to the achieved relative density (Fig. 13). Consequently, it was concluded that at moisture contents above 80% of optimum, the FBB was unstable and difficult to work, but once the instability was overcome, the achieved relatively density was still adequate. The moisture content of the FBB produced with the existing material was more variable than with the new crushed rock and the FBB workability was more sensitive to higher moisture contents. This likely reflects the increased plasticity associated with the fine particles in the natural gravel, as well as the more variable pre-production moisture content of the existing pavement materials compared to the quarried new crushed rock.

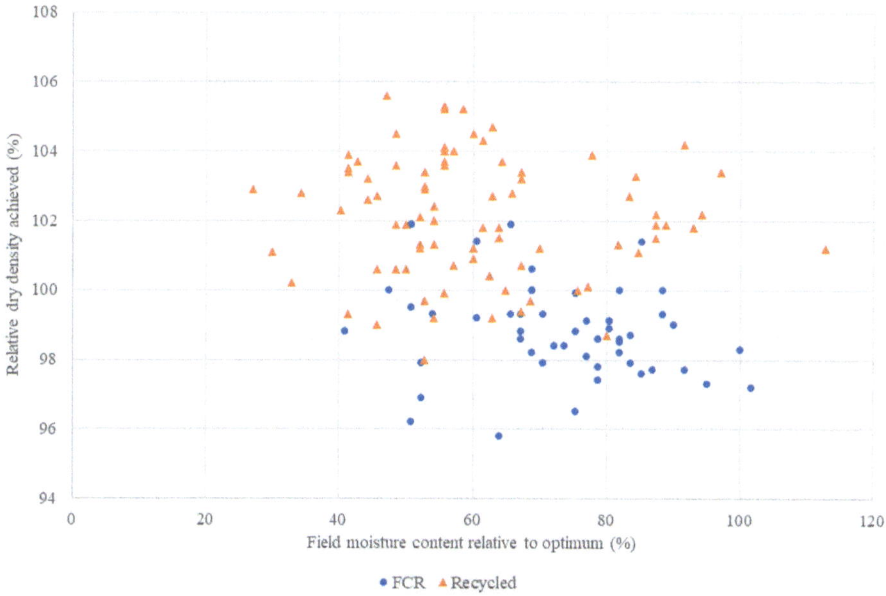

Fig. 13. Relative moisture content versus Relative density

7 Challenges

7.1 Representative Samples in Mixture Design

The recovery of representative samples of the existing pavement material during the FBB mixture design phase presented a significant challenge. The materials obtained were limited in volume and this restricted the number of mixture design iterations possible. Importantly, although the recovered materials were consistent in composition to the materials encountered during construction, the degree of pulverisation was not.

The inadequate pulverisation achieved by the expedient excavation of the existing asphalt surface using cold planing machines required secondary pulverisation in an on-site crusher and replacement of the lost volume of oversized asphalt pieces. Furthermore, the binding of the fine aggregate in the natural gravel by the freshly pulverised RAP required the addition of 20% sandy loam to achieve an appropriate grading for foamed bitumen stabilisation. Identifying these requirements during the on-site construction trials delayed the transition from new crushed rock to reuse of the existing pavement material by approximately one week.

7.2 Controlling Moisture

During construction is was identified that the FBB workability was sensitive to the production and field moisture content. Due to the more variable nature of the existing pavement material, controlling the moisture was more difficult than it was for the new crushed rock FBB. Over time, increased attention was placed on keeping the FBB moisture content at the lower end of the recommended 60–80% of optimum. However, the field moisture content remained variable and when it exceeded 80% of optimum, the FBB became unstable and difficult to work during compaction.

7.3 Latent Conditions

Isolated areas of unsuitable underlying pavement and subgrade material were encountered during construction. When located, the unsuitable material was removed and replaced by ballast over a geomembrane and finished with a thin layer of FBB to create a suitable working platform below the nominal 250 mm thick FBB layer. In the most extreme example, the excavation was 800 mm deep and this impacted overall project productivity and required some isolated rework where the surface was not adequately compacted prior to the end of the work period.

It is likely that an in-situ stabilisation process would have bridged over some of these isolated areas of unsuitable material. However, the identification and treatment of these areas effectively 'proved' the pavement and reduced the risk of isolated structural deficiencies requiring rectification in the future.

8 Conclusions

FBB is a valuable material for the expedient construction and reconstruction of pavements, particularly at airports with only one main runway and pavements that must be returned to a serviceable condition at the completion of each work period. The work at WCA demonstrated the benefits of FBB produced from both new and existing granular materials. FBB produced from new crushed rock was more consistent and was readily obtainable in a condition that was representative of the material encountered during construction for the purpose of mixture design. The existing pavement material was less predictable and this created a range of challenges. However, the cost and environmental benefits associated with reuse of existing pavement material in stabilised layers is significant and justifies the additional attention that is required during the design, production and construction processes. It is recommended that airports requiring pavement upgrades in short nighttime work periods consider FBB as a potential solution.

Acknowledgments. Although the views and opinions expressed are solely those of the author, the contribution to the project, collection of data and sharing of experiences by Whitsunday Coast Airport (the client) and Fulton Hogan (the construction contractor) are greatly appreciated and gratefully acknowledged.

References

Austroads: Review of Foamed Bitumen Stabilisation Mix Design Methods. Report AP-T178/11, Sydney, Australia, June 2011

Forsyth, P.: Airport infrastructure for the Airbus A380: cost recovery and pricing. J. Transp. Econ. Policy **39**(3), 341–362 (2005)

White, G.: Foamed bitumen stabilisation for Australian airports. In: Airfield Engineering and Maintenance Summit, Furama Riverfront, Singapore, 25–28 March 2014

White, G.: Towards a performance-based airport asphalt specification. In: International Conference on Highway Pavements and Airfield Technology, Philadelphia, Pennsylvania, USA, 27–30 August 2017

White, G.: Design and construct contracts for airport asphalt resurfacing. In: GeoChina 2018, HangZhou, China, 23–25 July 2018 (2018a)

White, G.: Foamed bitumen base for airport pavements. In: 28th Australian Roads Research Board International Conference, Brisbane, Queensland, Australia, 29 April–2 May 2018 (2018b)

Mechanistic Design of Foamed Bitumen Stabilised Pavement Bases

Sameera Pitawala, Arooran Sounthararajah(✉), and Jayantha Kodikara

Department of Civil Engineering, Monash University, Clayton Campus, Clayton, VIC 3800, Australia

{sameera.pitawala, arooran.sounthararajah, jayantha.kodikara}@monash.edu

Abstract. Foamed bitumen stabilised (FBS) materials as non-continuously bound pavement layers are being increasingly used globally in both new road pavement base construction and rehabilitation of degraded pavements. However, the essential research on testing and design methods of FBS materials is still far from complete. In current pavement design practice, the laboratory-derived fatigue criterion of asphalt is used directly for FBS materials. This paper critically discusses the limitations in current pavement design and test methods and proposes a detailed bottom-up study for the characterisation of the flexural fatigue behaviour of FBS materials under both laboratory and field conditions. The numerical results obtained from Circly computer program revealed that the critical pavement response under standard axle loading is below the inner wheel of the standard axle along the vehicle travel direction. The analysis of numerical results showed that the level of interaction between the axles of various axle configurations increases with increasing FBS base layer thickness. Moreover, the ratio of the maximum compressive stress divided by the maximum tensile stress (R) at bottom of the FBS base under standard axle loading was determined to be less than 10% for all the FBS base layer thicknesses employed in this study.

1 Introduction

As the demand increases for cost-effective and environmentally friendly pavement stabilisation, foamed bitumen stabilised (FBS) materials as non-continuously bound pavement layers are increasingly considered worldwide for road pavement construction. The resilient modulus of FBS materials is generally lower than that of hot mix asphalt (Loizos 2006), but can reach the resilient modulus of cement and lime treated materials (Sakr and Manke 1999). FBS materials are moisture sensitive and a reduction of resilient modulus between 30% and 50% have been reported when the specimen was soaked (Sakr and Manke 1999). Moisture sensitivity can be reduced through the addition of 1–2% hydrated lime or cement (Andrews 2002; Kendall et al. 1999; Saleh and Transfund New 2004). One of the key advantages of foamed bitumen stabilisation of pavement bases, in contrast to lime or cement stabilised pavement bases is that the FBS pavements can be trafficked almost immediately after their construction (Saleh

S. El-Badawy and R. Abd El-Hakim (Eds.): GeoMEast 2018, SUCI, pp. 112–123, 2019.
https://doi.org/10.1007/978-3-030-01908-2_9

2007). FBS pavements can also tolerate heavy rainfalls right after the construction (Kendall et al. 1999; Ramanujam and Jones 2000). Moreover, FBS materials have equivalent strength characteristics as cement treated materials while retaining flexibility, hence relatively high fatigue resistant.

Recent trials have demonstrated that fatigue cracking is one of the key distress mechanisms of FBS pavements (Austroads 2011a). Figure 1a displays the three phases of fatigue cracking for cementitious materials identified under traffic loading in the field (Theyse et al. 1996). Nonetheless, FBS materials exhibited only two phases of fatigue cracking, namely, fatigue life phase and equivalent granular phase, in the laboratory flexural fatigue tests reported by Dadras (2016) (see Fig. 1b). During the fatigue life phase, the stiffness of FBS material was reduced at a great rate from a value of 1,300 MPa to 800 MPa. However, it then decreased at a slow and stable rate in the equivalent granular phase. The laboratory stiffness variation of FBS materials under fatigue loading also agreed with the continuous falling weight deflectometer (FWD) measurements performed on the FBS pavements in service (Valentin et al. 2014). The initial flexural modulus of asphalt is normally defined as the flexural modulus of the beam specimen at the 50th load cycle. On the basis of the flexural modulus variation of pavement materials during repetitive loading, the flexural fatigue life of the cemented pavement materials is widely defined as the number of load cycles (N) required to reduce the flexural modulus of the beam specimens by 50% of their initial flexural modulus (Jameson 2010).

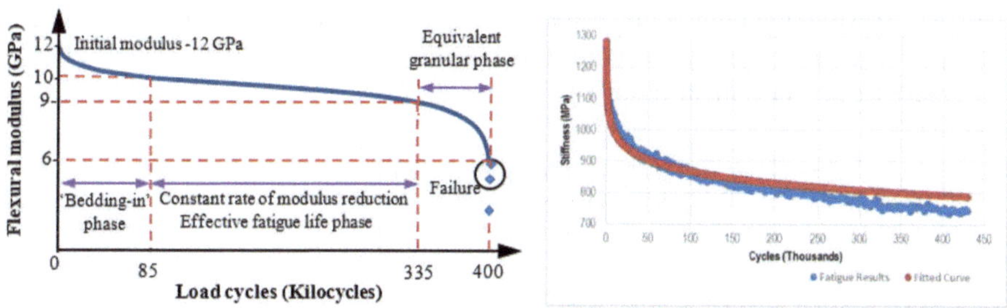

Fig. 1. (a) Typical flexural modulus variation during fatigue tests (Jameson 2010). (b) Flexural modulus variation for FBS (Dadras 2016)

Notwithstanding the increasing use of FBS materials in the road pavement construction, the design and test methods for FBS materials have not yet been developed scientifically that precludes their potential application in service. Though the laboratory tests do not simulate the actual field conditions, including the three-dimensional stress conditions and different axle load and configurations, by selecting appropriate laboratory test apparatus and test methods, the gap between laboratory and field conditions can be minimised. The four-point bending test is commonly used in the laboratory for the determination of flexural fatigue life of cement treated and asphalt pavements (Austroads 2011c). This test is considered to be a closer representation of the flexural stress/strain gradients which is applied to these pavements in the field (Austroads

2012). Dadras (2016) have suggested strain control with a sinusoidal cyclic loading to evaluate the flexural fatigue performance of FBS beams in the laboratory. However, the use of sinusoidal cyclic loading is appropriate when a direct comparison is made with European practice. In recently published standards, sinusoidal stress pulses have been used to determine the flexural fatigue performance of asphalt pavement bases (ASTM 2014), while the haversine stress pulses have been employed for cement pavement bases (Gonzalez et al. 2010). Thus, there appears to be a necessity for devising an appropriate laboratory test method for the characterisation of flexural fatigue performance of FBS materials.

With this background, the following sections will first critically analyse the current design approach for FBS pavements, and then evaluate the performance of FBS pavements under different loading conditions to develop an appropriate stress pulse shape for laboratory fatigue testing.

2 Current Design Practice

2.1 Fatigue Criteria of Various Pavement Materials

The relationships shown in Eqs. (1) and (2) are currently used in Australia to estimate the in-service fatigue life of the cement treated bases and asphalt pavement layers, respectively. However, both fatigue relationships were developed on the basis of the laboratory fatigue characteristics of the pavement materials. Besides, Eq. (2) is generally used Australia to predict the in-service fatigue life of the FBS material as there is no fatigue criterion currently exists for this material (Austroads 2012; Dadras 2016).

Cement treated pavement bases (Austroads 2012)

$$N = RF\left[\frac{\frac{113000}{E^{0.804}} + 191}{\mu\varepsilon}\right]^{m} \tag{1}$$

N = allowable number of repetitions of the load
$\mu\varepsilon$ = load-induced tensile strain at base of cement stabilised material (micro-strain)
E = cement stabilised material modulus (MPa)
m = load damage exponent
RF = reliability factor for cement stabilised fatigue

Asphalt pavement bases (Austroads 2012)

$$N = RF\left[\frac{6918(0.856V_b + 1.08)}{S_{mix}^{0.36}\mu\varepsilon}\right]^{5} \tag{2}$$

N = allowable number of repetitions of the load
$\mu\varepsilon$ = tensile strain produced by the load (microstrain)

V_b = percentage by volume of bitumen in the asphalt (%)
S_{mix} = asphalt modulus (MPa)
RF = reliability factor for asphalt fatigue

2.2 Axle Configurations

Figure 2 depicts the common axle configurations found in Australia. Table 1 lists the equivalent axle load of each axle configuration which is considered to cause an equivalent damage to the pavement structure as the standard axle. The 'standard axle' is defined as the single axle with dual tires (SADT), which transmits an axle load of 80 kN to the pavement structure. On the assumption that the equal deflection of pavement bases causes equal damage (Austroads 2008a), the equivalent axle loads were empirically derived for different axle configurations.

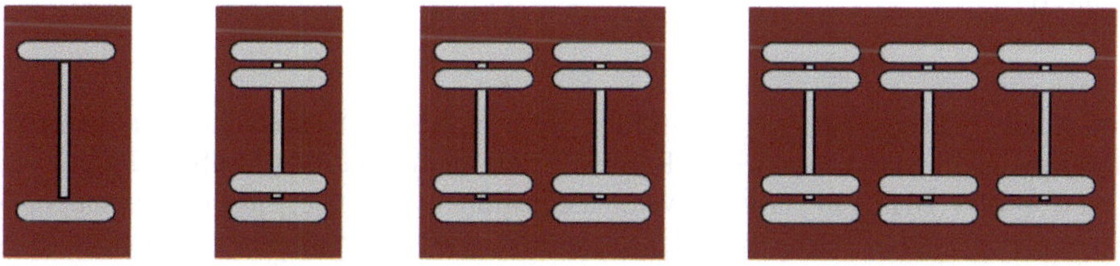

Fig. 2. Different axle configurations

The pavement damage due to various axle loads can be calculated using the equivalent axle loads given in Table 1. The following empirical equation (Eq. 3) is used to find the number of standard axle load repetitions required to produce the damage caused by any given axle load of a particular axle configuration (Austroads 2012).

$$SAR_m = \left(\frac{L_{ij}}{SLi}\right)^m \tag{3}$$

SAR is the number of standard axle repetitions which produce a same amount of damage as the equivalent axle load SL_i of axle configuration i (Table 1). L_{ij} is the jth axle load magnitude on the axle configuration i and m is the load damage exponent for the damage type. Currently, the load damage exponent (m) used in fatigue relationships (i.e. m = 12 for cement treated materials and m = 5 for asphalt) is attempted to use for determining the standard axle repetitions (SAR).

2.3 Mechanistic Design Procedure

In New Zealand and South Africa, FBS materials are assumed to be anisotropic ($E_{vertical} = 2 \times E_{horizontal}$) and linear elastic with a Poisson's ratio of nearly 0.3 for pavement design purposes (AshphaltAcademy 2009; Asphalt Academy 2002;

Table 1. Equivalent axle loads (Austroads 2012)

Axle group type	Load (kN)	Tire contact area radius (mm)	Tire contact pressure (MPa)
Single axle with single tyres (SAST)	53	92.10	0.994
Single axle with dual tyres (SADT)	80	92.10	0.750
Tandem axle with single tyres (TAST)	90	92.10	0.844
Tandem axle with dual tyres (TADT)	135	92.10	0.633
Tri-axle with dual tyres (TRDT)	181	92.10	0.566
Quad-axle with dual tyres (QADT)	221	92.10	0.518

Austroads 2011b). However, in Australia, these materials are treated as homogeneous, isotropic and linear elastic with a Poisson's ratio of around 0.4 (AustStab 2005; Leek 2011; QMRT 2017). In current pavement construction practice, most of the guidelines recommend 1–2% cement (or lime) to stabilise the primary binder (i.e. foamed bitumen) for the following reasons: (1) early-age strength, (2) as a modifier to reduce plasticity, (3) dispersive binder for foamed bitumen, and (4) anti-stripping agent (Jenkins and Ebels 2007). The viscoelastic behaviour of FBS materials highly depends on their bitumen content and secondary binder content (lime or cement). A bitumen content ranging from 2% to 4% is employed in current FBS pavement construction practice in Australia. The addition of cement/lime to FBS material may alter the viscoelastic characteristics of foamed bitumen. For instance, an addition of a high secondary binder to FBS material with a low bitumen content may bring a material which exhibits elastic characteristics under the repeated traffic loading.

The pavement response under various axle loads is computed by the linear elastic computer program CIRCLY (Mincad Systems 2012). The possible critical locations of flexible pavement structures in terms of horizontal strains under the standard axel loading are shown in Fig. 3. For the standard axle loading, the shape of the tire contact area is assumed to be circular and the stresses exhibit over the loaded area is assumed to be uniform. Based on the above assumptions, the standard axle loading (SADT) is considered as four uniformly loaded circular areas along the axle where the tire contact pressure is taken as 750 kPa (Austroads 2012). The cumulative damage factor (CDF) obtained from the mechanistic analysis is used to assess the structural capacity of the candidate pavement. CDF is the ratio between the number of standard axle load repetitions determined from the design traffic load distribution (TLD) and the total allowable number of load repetitions.

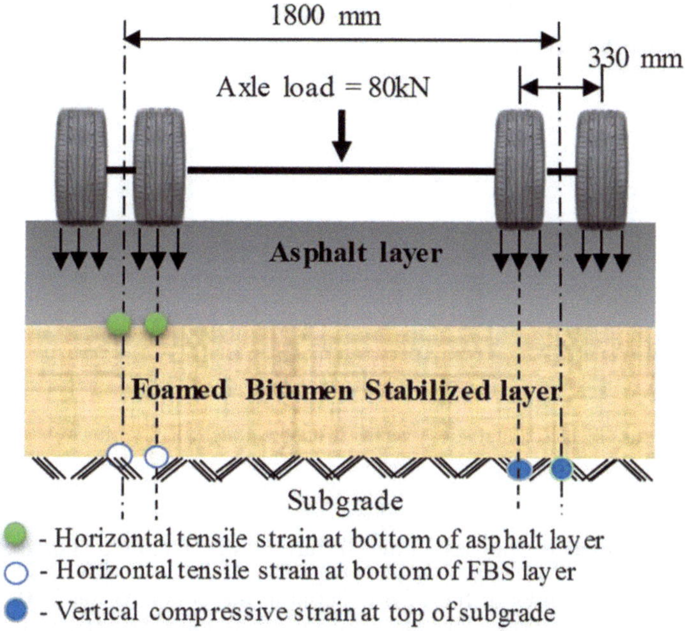

Fig. 3. Critical location of the stain in a pavement structure under standard axle loading (SADT)

3 Response of FBS Pavements Under Heavy Axle Loading

3.1 Selection of Parameters for Numerical Modelling of FBS Pavements

A pavement structure consists of asphalt layer (elastic modulus, E = 3,000 MPa and Poisson's ratio, v = 0.4), FBS base (elastic modulus, E = 2,500 MPa and Poisson's ratio, v = 0.4) and subgrade with a CBR of 6 was modelled in CIRCLY. All the layers were treated as elastic layers in accordance with current pavement design practice in Australia (Austroads 2011b). The elastic modulus and Poisson's ratio values of each material were chosen from the earlier studies (AustStab 2005; Dadras 2016; Valentin et al. 2014). A 50 mm thick asphalt layer and an FBS base layer with different layer thicknesses (100, 200 and 300 mm) were used in this analysis. The standard axle (Single Axle with Duel Tires -SADT) was used for the preliminary analysis of this research. The horizontal stress and strain variations at the bottom of the FBS base layer under standard axle along both longitudinal (Y) and transverse (X) directions were evaluated (see Fig. 4). The critical response of the pavement under standard axle loading is found to be the horizontal stress distribution at the bottom of the FBS base along the travel direction (σ_{yy}). Accordingly, σ_{yy} was then used for the rest of the analysis. The critical point along the transverse direction (X) in terms of stress and strain was then evaluated for different FBS base layer thicknesses (see Fig. 5). The results indicate that the critical point along the transverse direction is always closer to the inner wheel (i.e. X = 160 mm) of the half axle for all the FBS layer thicknesses used in this analysis. Therefore, the response of the FBS base layer immediately below the inner wheel (X = 160 mm) was chosen for the analysis presented in the following subsections.

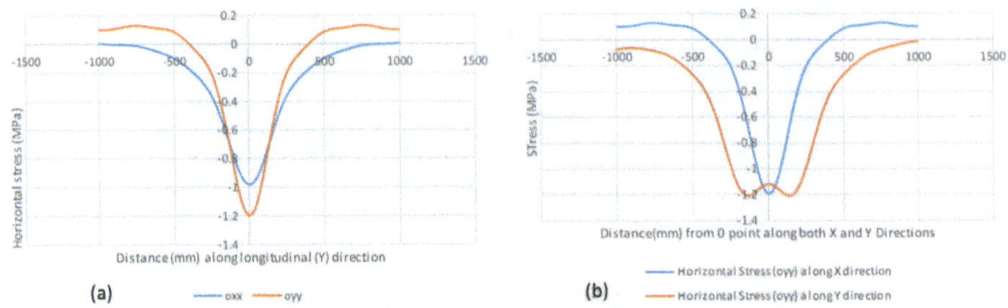

Fig. 4. Horizontal stress versus distance along x and y directions

3.2 FBS Pavement Responses Under Different Axle Configurations

The stress-strain responses of the FBS material at the bottom of the FBS base layer under standard axle (STDT) with a tire contact pressure of 0.75 MPa (equivalent axle load = 80 kN), tandem axle (TADT) with a tire contact pressure of 0.63 MPa (equivalent axle load = 135 kN) and tri-axle (TRDT) with tire contact pressure of 0.56 MPa (equivalent axle load = 180 kN) were computed along the longitudinal (Y) direction, respectively. For both the tandem axle and tri-axle, the level of interaction between the axles raises with increases in the FBS base thickness. The lowest level of interaction can be considered as two distinct loads as can be seen in Fig. 6.

3.3 Development of an Appropriate Stress Pulse Shape for Laboratory Testing

The foregoing section described the variation of critical stress at the bottom of the FBS base along the travel direction for various FBS base thicknesses. The stress shapes in distance domain were then translated in time domain using the vehicle speed as shown in the following relationship:

$$\text{Speed of the vehicle} = \text{t}\,\text{km/h} = \left[\text{t} \times 10^5/(3600 \times 1000)\right]\text{cm/ms} = (\text{t}/36)\,\text{cm/ms}$$

Note that when the wheel position is at Y = −1000 mm, the time was taken as zero for the plotting purposes.

The above relationship illustrates that when the vehicle moves with a speed of t km/h, each centimetre on the pavement surface has a time gap of (36/t) ms. Using this direct distance-time relationship, the stress pulse shapes were generated in the time domain for different vehicle speeds of different axle configurations (refer to Fig. 7). When a vehicle passed through a certain point on the pavement, the critical horizontal stresses at the bottom of the FBS base along the travel direction were determined. The

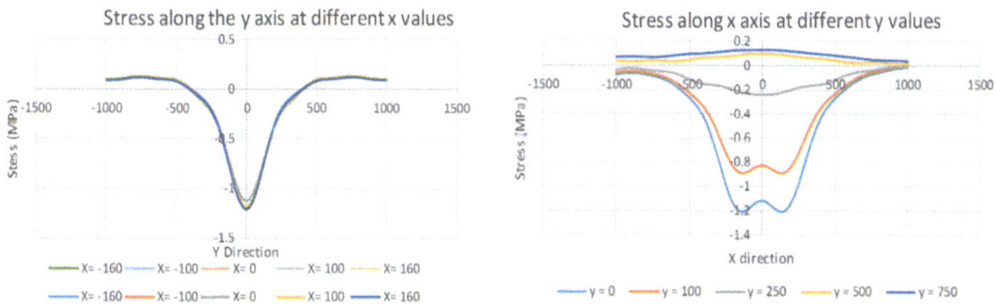

Fig. 5. Critical stress point identification along X and Y directions

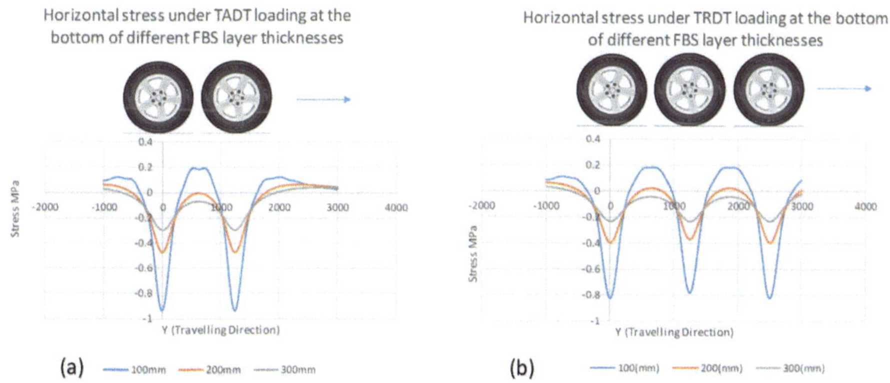

Fig. 6. Level of horizontal stress interaction with the thickness of FBS for different axle configurations

ratio between the maximum compressive and tensile stresses was then calculated for three different axle configurations and three different FBS thicknesses, respectively (see Fig. 8b). The width of the stress pulses (time) was then calculated for each axle configuration as shown in Fig. 9. As this figure shows, both the TADT and TRDT have the similar stress pulse width-layer thickness variation. However, for a given FBS base thickness, the stress pulse width for SADT is almost double when compared to the other axle configurations. In addition, the ratio between maximum compressive and tensile stresses (R) increases with increasing FBS base thickness, while the magnitude of both compressive and tensile stresses decreases. The maximum value of R is nearly 23% for TRDT axle configuration at an FBS base layer thickness of 100 mm. It is noted that for the standard axle configuration the R value is less than 10% for all the thicknesses of FBS base used in the present study. In addition, the difference between peak tensile and peak compressive stresses reduces with increasing number of axles and FBS base layer thickness as displayed in Fig. 8a.

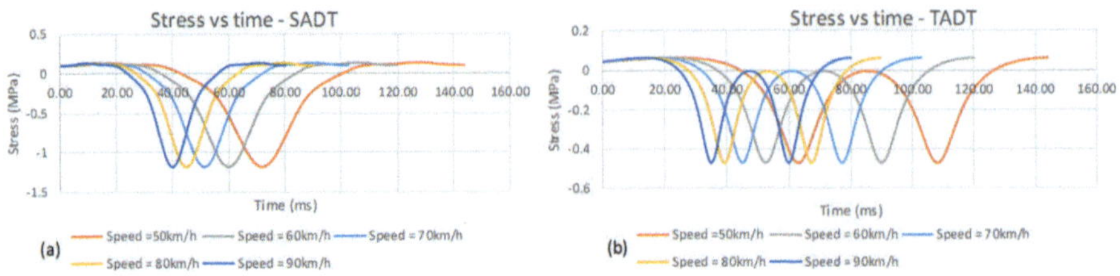

Fig. 7. Horizontal stress versus time relationship for different vehicle speeds

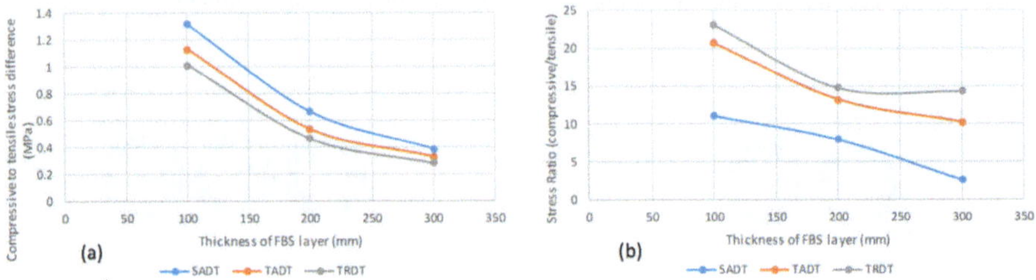

Fig. 8. Compressive and tensile stress behaviour of different axle configurations

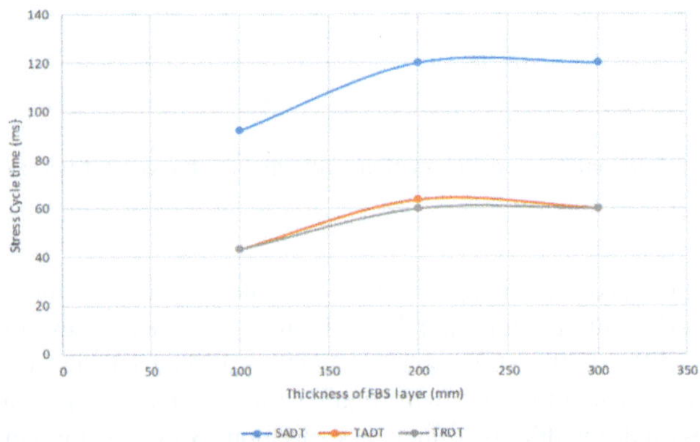

Fig. 9. Stress cycle time variation with the thickness of FBS layer for different axle configurations.

4 Issues in Current FBS Pavement Design Practice

Currently, there is no universally accepted laboratory derived fatigue relationship exists for FBS materials. Assuming that the performance of FBS material under repetitive heavy traffic loading is similar to asphalt, the asphalt fatigue criterion has been applied directly to FBS materials in pavement designs. A load damage exponent (*m*) of 5 is used in the asphalt fatigue relationship (Eq. 2); however, this value may not be appropriate for FBS materials. The load damage exponent values in the range between 5 and 20 have been reported for FBS materials (Dadras 2016). Also, a single value of *m* is currently used to convert the damage of any axle load of an axle configuration into an equivalent number of standard axle repetitions (Eq. 3). Though this approach is simple and convenient, more rigorous theoretical and experimental investigations are needed for successful pavement designs. Since the load damage exponent is incorporated as a power factor in the fatigue criterion, its incorrect choice can lead to major implications for the fatigue life of pavements in service.

The flexural modulus of bound materials, such as cement treated materials, is widely considered to be one of the design inputs for the mechanistic design of pavements (Sounthararajah et al. 2017). However, the loading conditions to determine the flexural modulus of bound materials are not specified in the standard modulus test method adopted in Australia (Arooran et al. 2015). Austroads (2011a) reported that the flexural modulus values obtained for cement treated materials decreased with increasing stress level. Incorrect choice of design modulus could also directly affect the performance of a pavement base in the field.

5 Conclusion and Recommendations

The numerical analysis using the Circly computer program revealed that the critical pavement response under standard axle loading is below the inner wheel of the standard axle along the vehicle travel direction (Y). Therefore, it is recommended that for the laboratory beam tests using four-point bending apparatus, an appropriate load pulse which simulates the critical response of the pavement should be considered. The analysis of numerical results showed that the level of interaction between axles for both TADT and TRDT increases with the increasing thickness of FBS base layers. As a result, the ratio between maximum compressive and maximum tensile stresses (R) increases with increasing thickness of FBS base layers, while the magnitude of both compressive and tensile stresses reduces. The maximum ratio (R) was reported as 23% under TRDT axle configuration at the minimum FBS layer thickness used in this study (i.e. 100 mm), and the ratio R was always less than 10% for standard axle configuration. Since the ratio between maximum compressive and maximum tensile stresses is negligible for standard axle configuration, the stress pulse that simulates the critical pavement response under standard axle is recommended for the laboratory flexural fatigue tests using four-point bending apparatus. Moreover, the frequency of the stress pulse, in other words, the loading duration of the stress pulse, was determined for various vehicle speeds and FBS base layer thicknesses.

Moreover, the damage evaluation of FBS pavements under the three-dimensional repetitive loading by a suitable material constitutive model should be captured, as the flexural fatigue experiments using four-point bending equipment do not simulate the exact field conditions. To develop an appropriate fatigue damage model, a series of laboratory tests can be conducted to characterise the microstructure of fatigue damage of FBS materials under different test conditions, including different curing ages and loading conditions. The model can then be implemented in PFC (particle flow code – a discrete element method software) and FBS bases could be modelled to predict their performance under various loading conditions in both laboratory and field. The shift factors associated with the change in boundary conditions from the laboratory to the field can then be evaluated for use in practical pavement designs.

Acknowledgment. This research project was partially Sponsored by the Australian Road Research Board (ARRB) and Nanocomm Hub at Monash university funded by the ARC ITRH Scheme. Their financial and in-kind support is gratefully acknowledged.

References

Andrews, N.: Foamed Bitumen Mix: Soil or Asphalt? (2002)

Arooran, S., Bui, H.H., Jitsangiam, P., Kodikara, J.: Development of new mechanistic pavement design approach for cement stabilized bases. Paper presented at the international conference on geotechnical engineering ICGE-Colombo-2015, Colombo, Sri Lanka (2015)

Ashphalt Academy: Technical guideline-bitumen stabilized materials. In: A Guideline for the Design and Construction of Bitumen Emulsion and Foamed Bitumen Stabilised Materials (2009)

Asphalt Academy: The Design and Use of Foamed Bitumen Treated Materials, Pretoria, South Africa (2002)

ASTM: Standard test method for determining fatigue failure of compacted asphalt concrete subjected to repeated flexural bending, West Conshohocken (2014)

Austroads: Technical basis of Austroads guide to pavement technology: Part 2: pavement structural design (2008a). (AP-T98-08), Austroads, Sydney

Austroads: A laboratory study of the influence of multiple axle loads on the performance of a cement treated material (2011a). (AP-T185-11:33-34). Austroads, Sydney

Austroads: Review of Foamed bitumen Stabilisation Mix Design Methods (2011b)

Austroads: Review of structural design procedures for foamed bitumen pavements (2011c). (AP-T188-11). Sydney

Austroads: Guide to pavement technology: part 2: Pavement structural design (2012). (AGPT02-12). Austroads, Sydney

AustStab: Design of foamed bitumen layers for roads, QLD (2005)

Dadras, E.: An Assessment of the Asphalt Fatigue Equation for Bitumen Stabilised Materials. (Master of Philosophy), Curteen University (2016)

Gonzalez, A., Howard, A., De Carteret, R.: Cost effective structural treatments for rural highways: cemented materials (2010)

Jameson, G.: Guide to Pavement Technology: Part 2: Pavement Structural Design (2010). (1921709154)

Jenkins, K., Ebels, J.: Mix design of bitumen stabilised materials: best practice and considerations for classification university. Paper presented at the Matieland proceedings of the 9th conference on asphalt pavements of Southern Africa (CAPSA) (2007)

Kendall, M., Baker, B., Evans, P., Ramanujam, J.: Foamed bitumen stabilisation. Paper presented at the proceedings roads at work-developing Southern Queensland, southern region symposium, QLD Department of Main Roads, Goondiwindi (1999)

Leek, C.: Review of the performance of in-situ foamed bitumen stabilised pavements in the City of Canning. Paper presented at the proceedings of the international conference on advances in geotechnical engineering (2011)

Loizos, A.: Estimation of Certain Foamed Asphalt Layer Properties of Cold in–Place Recycling (2006)

Mincad Systems.: CIRCLY 5: Software reference manual. In. Mincad Systems Pty Ltd, Richmond South, Victoria, Australia (2012)

QMRT.: Testing of Materials for Foamed Bitumen Stabilisation (150), Brisbane, Australia (2017)

Ramanujam, J., Jones, J.: Characterisation of foamed bitumen stabilization. Paper presented at the proceedings road system and engineering technology forum (2000)

Sakr, H.A., Manke, P.G.: Innovations in Oklahoma foamix design procedures (1999)

Saleh, M., Transfund New, Z.: Detailed Experimental Investigation for Foamed Bitumen Stabilisation. Transfund New Zealand, Wellington (2004)

Saleh, M.F.: Cost evaluation of foam bitumen and other stabilisation alternatives. Int. J. Pavement Eng. **8**(2), 157–161 (2007)

Sounthararajah, A., Wong, L., Nguyen, N., Bui, H.H., Kodikara, J.: Evaluation of flexural behaviour of cemented pavement material beams using distributed fibre optic sensors. Constr. Build. Mater. **156**, 965–975 (2017). https://doi.org/10.1016/j.conbuildmat.2017.09.027

Theyse, H., De Beer, M., Rust, F.: Overview of South African mechanistic pavement design method. Transp. Res. Rec. J. Transp. Res. Board **1539**, 6–17 (1996)

Valentin, J., Suda, J., Formanová, Z., Mollenhauer, K., Engels, M., Batista, F., Mcnally, C.: Introduction to European COREPASOL Project on Harmonizing Cold Recycling Pavement Techniques. Mater. Infrastruct. **2**, 1–17 (2014)

Laboratory Studies to Assess the Fatigue Behavior of Bituminous Concrete Mix

Lokesh Gupta[1(✉)] and Rakesh Kumar[2]

[1] Department of Civil Engineering, Sir Padampat Singhania University, Udaipur, India
lokeshgupta699@gmail.com
[2] Department of Civil Engineering, S.V. National Institute of Technology Surat, Gujarat, India

Abstract. Bituminous concrete is a premix widely used in road construction as a surface course. A high amount of expenditure is required for the construction of pavement structure. Due to increase in the magnitude of commercial vehicles, bituminous mixes are now exposed to greater stresses, and hence the performance of bituminous mixes is not good enough for paving applications. The present investigation, primly emphasizes to study the effect of filler materials and stress levels based on tensile strength on the number of load repetition sustained against the initiation of the first crack of a bituminous concrete mix. At controlled test temperature 25 °C, Indirect tensile fatigue test is conducted on bituminous concrete mix prepared using Stone dust and Cement as filler materials at 10%, 20%, and 30% stress level. Further regression analysis Based on the data obtained from Indirect Tensile Fatigue test and cost estimation per km for the construction of one kilometer of bituminous concrete mix prepared using both types of filler materials is carried out. As per the limited laboratory studies carried out, It can be concluded that no of load repetition varies with the addition of filler content and filler type. Fatigue life against the cracks decreases with an increase in stress level irrespective of filler materials. In the mathematical model Stress level and tensile stress are statistically significant with fatigue life, i.e., no of load repetition of a bituminous concrete mix.

Keywords: Fatigue life · Stress · Indirect tensile fatigue test · Stone dust Cement

1 Introduction

Presently rise in traffic flow has been noticed on most of the road stretches in India. Bituminous layer within the pavement structure experiencing a high amount of stress caused subsequently due to increases axle load, contact and tyre pressure. In addition to this considerable change in pavement temperature and stress developed may further result in progress of pavement distress. Appropriate selection of filler type and its share in the bituminous mix improves performance and fatigue behaviour of surface course (Gupta and Suresh 2017).

Bituminous concrete is strong enough to handle years of vehicle traffic and is relatively easy to repair or refinish (Rodrigues 2000). Fatigue failure is one of the main

S. El-Badawy and R. Abd El-Hakim (Eds.): GeoMEast 2018, SUCI, pp. 124–134, 2019.
https://doi.org/10.1007/978-3-030-01908-2_10

distress mechanisms causing degradation of pavements. Fatigue is caused by repeated traffic loadings, which result in a crack initiation, crack propagation and eventually catastrophic failure of the material due to unstable crack growth (Jacobs 1995).

Large variations in seasonal and daily temperatures are responsible for the early deterioration of flexible pavements (Shukla 1987). During the summer, high temperature can soften the bitumen binder and consequently reduce the stiffness of bituminous mixes. On the other hand, in winter, low temperatures can stiffen the binder and reduce the flexibility of paving mix resulting in low temperature cracking.

The bitumen binder obtained from refineries meet the existing specifications but does not provide the desired performance in the field. Some of the studies have shown that incorporation of additives can improve the properties of bitumen or blend of additives such as sulfur, polymers, rubber etc (Sikdar et al. 1999; Gupta and Suresh 2016; Gupta et al. 2016; Gupta and Bellary 2018; Shukla and Jain 1989).

Key fatigue model is broadly classified into four criteria, i.e., Laboratory study based (Ghuzlan and Carpenter 2003); dissipated energy criteria based (Aglan and Figueroa 1993; Bonnetti et al. 2002; Si et al. 2002; Daniel and Bisirri 2005; Carpenter and Shen 2006); models based on fracture mechanics (Ghuzlan and Carpenter 2006; Abdulshafi and Majidzadeh 1985; Majidzadeh et al. 1972; Mull et al. 2002; Sulaiman and Stock 1995; Button et al. 1987; Ramsamooj 1999); and models based on damage mechanics (Lee et al. 2000; Bodin et al. 2004; Suo and Wong 2009; El-Basyouny and Witczak 2005; Castro and Sanchez 2007; Rodrigues 2000). American Association of State Highway and Transportation Officials (AASHTO) suggested repeated flexural bending test to determine the fatigue performance of bituminous mix (Wen and Bahia 2009; Huang et al. 2007; Epps et al. 1969; Tigdemir et al. 2002; Monismith et al. 1961; Sousa et al. 1993).

Read (1996) used ITT to study the fatigue behavior on bituminous mixes. Fatigue studies were done for four mixes consisting of 30/40 grade Hot Rolled Asphalt (HRA), Styrene Butadiene Styrene (SBS) and 20 mm dense bituminous macadam (DBM).

The conventional literature explains the mechanism of development of fatigue cracks. Continuous application of wheel load causes higher stresses initiate fatigue cracks at the bottom fibre layer and subsequently propagates towards the top fibre (Pell 1973). On the other hand, a study has represented that cracks may occur in the top fibre and propagates downwards. Whether it is top–down cracking or bottom–top cracking, fatigue cracking required the critical attention of highway researchers to develop the reliable technique of corrections (Nunn 2014).

2 Objective of Present Study

The objectives of the present study are

- To assess the fatigue behaviour of bituminous concrete mix prepared using cement and stone dust as filler materials.
- To establish the prediction equation of fatigue life using regression analysis.

3 Material Characterization

Coarse aggregates, fine aggregates, mineral filler and binder, are the general constituents of a bituminous mix. In the present study VG-30 as a binder while Stone dust and cement are used as mineral filler.

3.1 Bitumen

Bitumen coats the aggregate as well as fills the part of a volume of air pockets in between aggregates. The suitability of (VG 30) bitumen as a binder material, tests were conducted in the laboratory. The test results satisfy the requirements as per IS 73:2013 and are presented in Table 1. The specific gravity of bitumen (VG-30) is 1.01.

Table 1. Test results of binder (VG-30)

Tests particulars	Test results	Requirement as per IS73:2013
Penetration at 25 °C, 100 gm, 5 s, 0.1 mm	63	Min. 45
Softening point (Ring & Ball), °C	51.7	Min. 47
Flash point, °C	273	Min. 220
Fire point, °C	291	Min. 220
Ductility, cm	89	Min. 75
Specific gravity	1.01	Min. 0.99
Kinematic viscosity @135 °C, CP	466	Min. 350

3.2 Aggregates

Aggregates mainly consisting of both coarse and fine aggregates (coarse aggregate of 13.2 mm to 2.36 mm and fine aggregates of 2.36 mm to 75 μ) were used. Coarse aggregate imparts better shear and compressive strength and hence enhance the fine aggregate fills the interlocking characteristics and volume of air pockets between the coarse aggregates. To assess the properties of aggregate as per MORT&H (Vth Revision 2013) specification, tests are conducted in the laboratory, and the test results are presented in Table 2. Aggregate gradation for the Bituminous concrete mix (Grading-2) is adopted for the present study as per MORT&H (Vth Revision 2013) and shown in Table 3.

3.3 Mineral Filler

Mineral filler fills the voids, stiffens the binder and offers permeability. In the present work stone dust and cement are used as mineral filler. A specific gravity of stone dust and cement are 2.60 and 3.09 respectively. The gradation requirement was tested in the laboratory and results obtained are presented in Table 4.

Table 2. Test results of aggregates

Test particulars	Test results	Requirements as per MORT&H (Vth Revision)
Aggregate impact value (%)	20.50%	Max. 24%
Los Angeles abrasion value (%)	22.15%	Max. 30%
Flakiness and elongation index (combined), (%)	26.57%	Max. 30%
Aggregate stone polishing value	63	Min. 55
Water absorption (%)	0.23%	Max. 2%
Aggregate specific gravity		
Coarse aggregate	2.66	Min. 2.5
Fine aggregate	2.64	

Table 3. Aggregate gradation for bituminous concrete mix

Sieve size in mm	% Passing (specified)	Obtained gradation
19	100	100
13.2	79–100	89.5
9.5	70–88	79
4.75	53–71	62
2.36	42–58	50
1.18	34–48	41
0.6	26–38	32
0.3	18–28	23
0.15	12–20	16
0.075	04–10	7

Table 4. Gradation details of filler material

IS sieve size (μ)	Cumulative % passing by weight of total aggregates		
	Obtained		Requirement as per MORT&H (Vth revision)
	Cement	Stone dust	
600	100	100	100
300	100	100	95–100
150	95.80	90.32	85–100

4 Test Programme

4.1 Indirect Tensile Fatigue Test (ITFT)

Indirect Tensile Fatigue Test was performed at 25 °C temperature to assess the impact of filler materials on the fatigue behaviour of Bituminous Concrete Mix (Grade-2) prepared using stone dust and cement as filler materials, by varying stress levels i.e.,

10%, 20% and 30%. The stress levels were fixed based on the Indirect Tensile Strength Test conducted on cylindrical specimens of a bituminous concrete mix (grade-2).

The equipment has a facility to apply the repetitions in half sine waveform. The recording job of the applied load, deflection of both horizontal and vertical LVDT's for every cycle and the total number of repetitions before failure for each test was done by a data acquisition system with comprises of a computer. The frequency of loading and rest period can also be fixed as per the requirement. In this present investigation, the loading frequency adopted was 2 Hz (2 cycles per second) and the rest period was 0.2 seconds. A specimen was declared failed if the deflection has reached 5 mm on either of horizontal or vertical LVDT (whichever is earlier).

5 Results and Discussion

Indirect Tensile Fatigue test is conducted on a bituminous concrete mix (Grading-2) prepared using stone dust (1%) and cement (1%) as filler materials at optimum bitumen content. Specimens are conditioned at 25 °C and properties such as resilient modulus, initial tensile strain and fatigue life are determined at three different stress level i.e., 10%, 20% and 30%. The test results are presented in Table 5. The test programme was adopted because of its simplicity and performed as per ASTM D 4123.

Table 5. Fatigue life of bituminous concrete Mix

Stress level, %	Tensile stress, MPA		Fatigue life, No. of cycles		% Increase in fatigue life
	Stone dust	Cement	Stone dust	Cement	
10	0.0887	0.1	11832	14162	16
20	0.1751	0.196	3176	4288	26
30	0.263	0.296	924	1168	21

5.1 Effect of Stress on Fatigue Life

It can clearly attest to the Table 5 and Figs. 1, 2, that filler material has a significant impact on the fatigue behaviour of the bituminous concrete mix and stress is inversely associated with fatigue life of the mix. There is a reduction in the fatigue life of bituminous concrete mix irrespective of a type of filler material.

At 25 °C test temperature, fatigue life of bituminous concrete mix prepared using cement at three different stress level 10%, 20% & 30% is 14162, 4288, 1168 while it is 11832, 3176 and 924 for BC mix prepared using stone dust.

There is a substantial increase in the fatigue life of bituminous concrete mix prepared using cement as filler material of about 16%, 26% and 21% respectively when compared to bituminous concrete mix prepared using stone dust as filler material. This can be attributed to the mix prepared using cement as filler material possess fewer air voids as compared to mix prepare using stone dust as filler material. Cement as filler material improves the aggregate interlocking by developing the better skeleton of aggregates within the mix and hence makes the mix relatively stiffer.

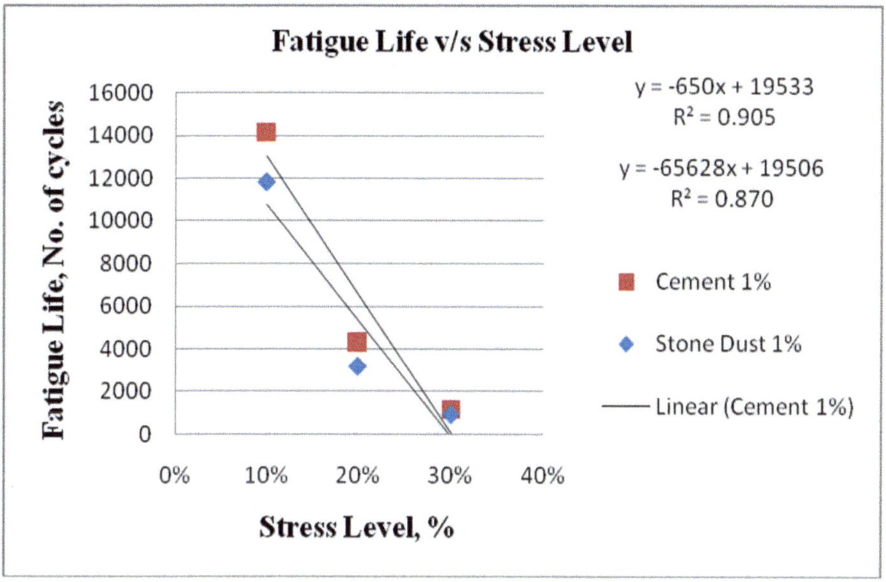

Fig. 1. Fatigue Life v/s Stress Level

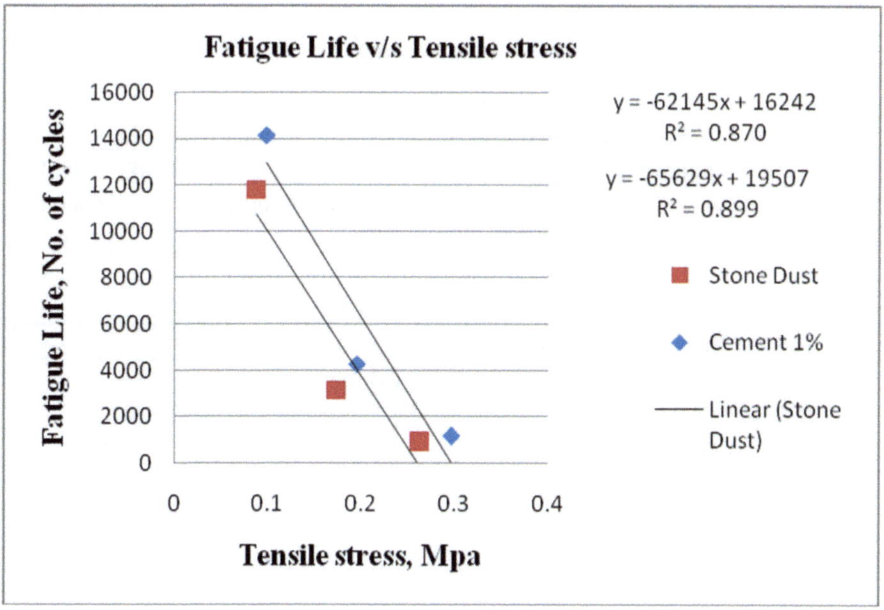

Fig. 2. Fatigue Life v/s Tensile stress

5.2 Cost Estimation

Cost required per km for the construction of Bituminous concrete mix prepared using Viscosity grade (VG-30) bitumen for 1 km road stretch considering width of 7 m. The estimation was prepared as per Scheduled Rates 2016 (SR 2016), PWD Jaipur circle. Cost estimation as presented in Table 8, the overall cost required per km for the construction of bituminous concrete layer/wearing course prepared using cement as the filler material is Rupees fifty two lakhs sixty six thousand five hundred sixty six only

The overall cost required per km for the construction of bituminous concrete layer/wearing course prepared using stone dust as the filler material is Rupees fifty one lakhs fifty four thousand thirteen only.

Table 6. Summary of fatigue life prediction equation (Stone dust)

Tensile Stress									
Model	**R**	**R Square**	**Adjusted R Square**	**Std. Error of the Estimate**	**Change Statistics**				
					R Square Change	**F Change**	**df1**	**df2**	**Sig. F Change (p)**
1	0.950	0.903	0.899	1801.43	0.903	232.60	1	25	0.000
a. Predictors: (Constant), Tensile Stress									
b. Dependent Variable: Fatigue Life									
FL_C = 19507 - 65629 TS									

Stress Level									
Model	**R**	**R Square**	**Adjusted R Square**	**Std. Error of the Estimate**	**Change Statistics**				
					R Square Change	**F Change**	**df1**	**df2**	**Sig. F Change (p)**
1	0.953	0.909	0.905	1745.65	0.909	249.33	1	25	0.000
a. Predictors: (Constant), Stress Level									
b. Dependent Variable: Fatigue Life									
FL_C = 19533 - 650 SL									

Table 7. Summary of fatigue life prediction equation (cement)

Tensile Stress									
Model	**R**	**R Square**	**Adjusted R Square**	**Std. Error of the Estimate**	**Change Statistics**				
					R Square Change	**F Change**	**df1**	**df2**	**Sig. F Change (p)**
1	0.939	0.882	0.877	1691.19	0.882	186.144	1	25	0.000
a. Predictors: (Constant), Tensile Stress									
b. Dependent Variable: Fatigue Life									
FL_S = 16242- 62145 TS									
Stress Level									
Model	**R**	**R Square**	**Adjusted R Square**	**Std. Error of the Estimate**	**Change Statistics**				
					R Square Change	**F Change**	**df1**	**df2**	**Sig. F Change (p)**
1	0.942	0.887	0.870	1805.67	0.887	54.73	1	25	0.000
a. Predictors: (Constant), Stress Level									
b. Dependent Variable: Fatigue Life									
FL_S = 19506 - 65628 SL									

Table 8. Cost required per km for the construction of bituminous concrete

S. No.	Description of items	Rate (Rs.)	
		Cement 2%	Stone dust 2%
1	Materials	4,356,608.0	425,4055.0
2	Machineries	497,000.00	497,000.00
3	Consumables	274,778.00	274,778.00
4	Crew charges	4900.00	4900.00
5	Labours	129,080.00	129,080.00
6	Small T&P, Miscel	4200.00	4200.00
Total cost of mix per km		5,266,566.00	5,164,013.00

6 Mathematical Model

The linear regression analysis has been done using SPSS software to develop the fatigue life prediction equation of bituminous concrete mix prepared using cement and stone dust as filler material, Fatigue life prediction equation is developed in the following form

$$FL_K = a + bSL \tag{1}$$

$$FL_K = a + bTS \tag{2}$$

where FL_K = Fatigue Life of a bituminous concrete mix, K = Filler Material, Stone dust or Cement, a = Constant, b = Regression Coefficient, SL = Independent Variable i.e., Stress Level, TS = Independent Variable i.e., Tensile Stress

To analyse the data, SPSS software is used with linear regression method. The method is opted to predict the number of load repetition, i.e., fatigue life of bituminous concrete mix considering the stress level and tensile stress as an independent variable.

6.1 Prediction Equation for Stone Dust

Details of fatigue life prediction equation of bituminous concrete mix prepared using stone dust are presented in Table 6. Prediction equations as

$$FL_S = 16242 - 62145\ TS \tag{3}$$

$$FL_S = 19506 - 65628\ SL \tag{4}$$

Independent variable opted, i.e., tensile stress, and stress level are explaining 87.7% and 87% of accuracy (Adjusted R^2) respectively. Further, it can also attribute that tensile stress, and stress level are significantly associated with the fatigue life, i.e., no of load repetition of a bituminous concrete mix as the regression coefficient plus constant are significant at 0.05 levels.

6.2 Prediction Equation for Cement

Details of fatigue life prediction equation of bituminous concrete mix prepared using stone dust are presented in Table 7. Prediction equation is as

$$FL_C = 19507 - 65629\ TS \tag{5}$$

$$FL_C = 19533 - 650\ SL \tag{6}$$

Independent variable opted, i.e., tensile stress, and stress level are explaining 89.9% and 90.5% of accuracy (Adjusted R^2) respectively. Further, it can also attribute that tensile stress, and stress level are significantly associated with the fatigue life, i.e., no of load repetition of a bituminous concrete mix as the regression coefficient plus constant are significant at 0.05 levels.

7 Conclusions

By limited laboratory study carried out in the present study, the following conclusion is drawn

- There is substantial increase in the fatigue life of bituminous concrete mix prepared using cement as filler material as compared to mix prepared using stone dust as filler material.
- Bituminous concrete blend prepared using cement as filler material has better fatigue life than the mix prepared using stone dust as filler material.
- The reduction in fatigue life of bituminous concrete mix with the increase in stress level and stress irrespective of the type of filler material.
- Since there is the marginal difference in the cost of construction, it is concluded that either of filler material considered for this study can be used in road construction depending upon the availability.
- In the fatigue life prediction equation, stress level and stress are statistically significantly associated with the fatigue life of bituminous concrete mix.
- Based on the studies carried out, it can be concluded that the behaviour of bituminous concrete mix prepared using cement (2%) as the filler material is superior regarding Fatigue life.

References

Abdulshafi, A.A., Majidzadeh, K.: J-integral and cyclic plasticity approach to fatigue and fracture of asphaltic mixtures. Transp. Res. Rec. **1034**, 112–123 (1985)

Aglan, H.A., Figueroa, J.L.: Damage-evolution approach to fatigue cracking in pavements. J. Eng. Mech. **119**(6), 1243–1259 (1993)

Bonnetti, K., Nam, K., Bahia, H.: Measuring and defining fatigue behavior of asphalt binders. Transp. Res. Rec. **1810**, 33–43 (2002)

Button, J.W., Little, D.N., Kim, Y., Ahmed, J.: Mechanistic evaluation of selected asphalt additives. In: Proceedings of Association of Asphalt Paving Technologists, vol. 56, pp. 62–90. Association of Asphalt Paving Technologists (AAPT), St. Paul (1987)

Bodin, D., Pijaudier-Cabot, G., de La Roche, C., Piau, J.M., Chabot, A.: Continuum damage approach to asphalt concrete fatigue modeling. J. Eng. Mech. **130**(6), 700–708 (2004)

Carpenter, S., Shen, S.: Dissipated energy approach to study hot-mix asphalt healing in fatigue. Transp. Res. Rec. **1970**, 178–185 (2006)

Castro, M., Sanchez, J.A.: Damage based model for prediction of asphalt concrete fatigue curves. J. Mater. Civ. Eng. **19**(8), 700–702 (2007)

Daniel, J.S., Bisirri, W.M.: Characterizing fatigue in pavement materials using a dissipated energy parameter. In: Advances in Pavement Engineering, pp. 1–10 (2005)

El-Basyouny, M., Witczak, M.: Part 2: Flexible pavements: Calibration of alligator fatigue cracking model for 2002 design guide. Transp. Res. Rec. **1919**, 76–86 (2005)

Epps, J.A., Monismith, C.L., Warden, W.B., Pell, P.S., Kallas, B.F., Terrell, R.L., Mcleod, N.W.: Influence of mixture variables on the flexural fatigue properties of asphalt concrete. In: Association of Asphalt Paving Technologists Proceedings (1969)

Gupta, L., Suresh, G.: Effect of using cement and fly ash as filler materials on the marshall properties of bituminous concrete mix. Indian Highways **44**(4) (2016)

Ghuzlan, K.A., Carpenter, S.H.: Fatigue damage analysis in asphalt concrete mixtures using the dissipated energy approach. Can. J. Civ. Eng. **33**(7), 890–901 (2006)

Gupta, L., Patil, A., Ojha, A.: A study on the marshall properties Of DBM mix prepared using Vg-30 and Crmb-55 as binder materials. IJRET **5**(3), 32–36 (2016)

Ghuzlan, K.A., Carpenter, S.H.: Traditional fatigue analysis of asphalt concrete mixtures. In: TRB 2003 Annual Meeting. Recuperado el, vol. 3 (2003)

Gupta, L., Suresh, G.: Determination of indirect tensile strength of bituminous concrete mix prepared using stone dust and cement as filler materials. In: 'International Congress and Exhibition' Sustainable Civil Infrastructures: Innovative Infrastructure Geotechnology, pp. 249–261. Springer, Cham (2017)

Gupta, L., Bellary, A.: Comparative study on the behavior of bituminous concrete mix and warm mix asphalt prepared using lime and Zycotherm as additive. Mater. Today Proc. **5**(1), 2074–2081 (2018)

Huang, C., Najjar, Y.M., Romanoschi, S.A.: Predicting asphalt concrete fatigue life using artificial neural network approach (No. 07-1607) (2007)

Jacobs, M.M.J.: Crack growth in asphaltic mixes. Ph.D thesis, Delft University of Technology, Netherlands (1995)

Lee, H.J., Daniel, J.S., Kim, Y.R.: Continuum damage mechanics-based fatigue model of asphalt concrete. J. Mater. Civ. Eng. **12**(2), 105–112 (2000)

Majidzadeh, K., Kauffmann, E.M., Saraf, C.L.: Analysis of fatigue of paving mixtures from the fracture mechanics viewpoint. In: Fatigue of Compacted Bituminous Aggregate Mixtures. ASTM International, West Conshohocken (1972)

Mull, M.A., Stuart, K., Yehia, A.: Fracture resistance characterization of chemically modified crumb rubber asphalt pavement. J. Mater. Sci. **37**(3), 557–566 (2002)

Monismith, C.L., Secor, K.E., Blackmer, W.: Asphalt mixture behavior in repeated flexure. In: Proceedings of Association of Asphalt Paving Technologists, vol. 30, pp. 188–222 (1961)

Nunn, M.E.: A question of fatigue? In: Performance and Durability of Bituminous Materials, pp. 53–62. CRC Press, Cambridge (2014)

Pell, P.S.: Characterisation of fatigue behaviour. In structural design of asphalt concrete pavements to prevent fatigue cracking. In Proceeding of Symposium, Washington, Highway Research Board Special Report 140, Washington, pp. 49–64 (1973)

Ramsamooj, D.V.: Prediction of fatigue performance of asphalt concrete mixes. J. Test. Eval. **27**(5), 343–348 (1999)

Rodrigues, R.M.: A model for fatigue cracking prediction of asphalt pavements based on mixture bonding energy. Int. J. Pavement Eng. **1**(2), 133–149 (2000)

Read, J.M.: Fatigue cracking of bituminous paving mixtures. Doctoral dissertation, University of Nottingham (1996)

Shukla, R.S.: Choice of different paving grade bitumens for different climatic regions of India. Road Research Papers (N218) (1987)

Sikdar, P.K., Jain, S.S., Bose, S., Kumar, P.: Premature cracking of flexible pavements. J. Indian Roads Congr. **60**(3), 355–398 (1999)

Shukla, R.S., Jain, P.K.: Improvement of waxy bitumens by the addition of synthetic rubbers, polymers and resins. Highway Res. Bull. **38** (1989)

Sousa, J.B., Tayebali, A.A., Harvey, J.T., Hendricks, P., Monismith, C.L.: Sensitivity of SHRP-A003A testing equipment to mix design parameters for permanent deformation and fatigue. In: Annual Meeting of the Transportation Research Board (1993)

Sulaiman, S.J., Stock, A.F.: The use of fracture mechanics for the evaluation of asphalt mixes (with discussion). J. Assoc. Asphalt Paving Technol. **64** (1995)

Suo, Z., Wong, W.G.: Analysis of fatigue crack growth behavior in asphalt concrete material in wearing course. Constr. Build. Mater. **23**(1), 462–468 (2009)

Si, Z., Little, D.N., Lytton, R.L.: Characterization of microdamage and healing of asphalt concrete mixtures. J. Mater. Civ. Eng. **14**(6), 461–470 (2002)

Tigdemir, M., Karasahin, M., Sen, Z.: Investigation of fatigue behaviour of asphalt concrete pavements with fuzzy-logic approach. Int. J. Fatigue **24**(8), 903–910 (2002)

Wen, H., Bahia, H.: Characterizing fatigue of asphalt binders with viscoelastic continuum damage mechanics. Transp. Res. Rec. **2126**, 55–62 (2009)

Automatic Recognition of Pavement Degradation: Case of Rif Chain

Soukaina Meziane(✉), Lahcen Bahi(✉), and Latifa Ouadif(✉)

L3GIE Laboratory, Mohammadia Engineering School,
Mohammed V University, Rabat, Morocco
soukaina.meziane@gmail.com, ouadif@gmail.com,
lahcen.bahi@emi.ac.ma

Abstract. Rif Chain (North of Morocco) is subject of ground instabilities. Several studies analyzed landslides' triggering factors and confirmed that water is a predominant factor. Once a landslide occurs, it damages the roadway paralyzing traffic especially Mediterranean Bypass Road which links the North of Morocco from East to West. To open up population, recovery maintenance requires a heavy budget. Thus, this paper proposes a preventive solution based on automatic recognition of pavement degradations (cracking and tearing) that allow water penetration through pavement until reaching ground support or even deeper layers destabilizing slopes and then triggering landslides. This study applies deep convolutional neural network using pretrained AlexNet model, to automate image classification of road pavement and recognition of three classes (presence of cracking, presence of tearing and absence of degradation). Then to save more time in training and classifying, a second classifier is trained to simply distinguish between two classes (pavement degraded and pavement not degraded).

1 Introduction

Rif Chain is frequently subject to landslides (Maurer 1964; Chalouan 1986; El Fellah 1994; Fares 1994; El Khattabi 2001; El Khattabi and Carlier 2004; Sossey Alaoui 2005; Benmakhlouf et al. 2012; Ouadif et al. 2012; Baba et al. 2012; Mastere 2011; Mastere et al. 2015; El Fahchouch et al. 2015; Boujaaj et al. 2016; El-Fengour 2016). In particular, the Mediterranean Bypass Road contains dominated instability sections conditioning the state of pavement. Roads are degraded due to several factors such as traffic (Internal fatigue related to vehicles passage and fatigue of the wearing course related to the tires friction) and the climate (precipitation causes storm water runoff and ground water infiltration). Thus, water enters by pavement degradations via cracks or/and tearing. If road drainage network is deficient, water seeps sideways via roadway shoulders and ditches. In addition, it infiltrates by capillarity from groundwater. As a result, bearing capacity of soil decreases causing important pavement degradation. The terrain is geotechnical vulnerable, so presence of water influences its stability. Thereby, it is important to check pavement impermeability to prevent water penetration over it. For pavement protection, it is crucial to maintain shoulders and ditches (for rapid drainage of runoff and for avoiding bilateral penetration); and to lower water level in

S. El-Badawy and R. Abd El-Hakim (Eds.): GeoMEast 2018, SUCI, pp. 135–144, 2019.
https://doi.org/10.1007/978-3-030-01908-2_11

soil (by checking ditch slopes and installing drains). This article focuses on pavement degradation where water seeps and consequently may pose danger for population and road users. In this case, deep convolutional neural network DCNN is efficient for automatic recognition of pavement degradations (cracking and tearing) especially that nowadays, visual record is done manually. The roads linear is important and the collection of pavement images is done every 100 m. For example, Mediterranean Bypass counts more than 507 km. So 5070 images are taken for each pavement quoted right and left side. In fact, 10140 is the total images on the entire platform. Thus, DCNN classifies fast visual record for programming maintenance. More pavement is degraded more maintenance cost is higher.

2 Related Work

Automatic recognition of cracks is established using intensity-thresholding methods that detect cracks in pavement images (Zou et al. 2012). In fact, CrackTree evaluates with high performance road condition but requires sophisticated and expensive algorithms. Laser scanning methods present also high accuracy but it is expensive (Vijayan and Geethalakshmi 2018; Laurent et al. 2010). On the other hand, deep convolutional neural network is successfully used in several domains such as computer vision, object classification, automatic image counting, automatic speech and video recognition. It can be trained from scratch or even use pre-trained networks (transfer learning). Several pre-trained DCNN (AlexNet, VGG-16, VGG-19, ResNet, and Inception-V3) are available and give fast and accurate classification. Krizhevsky et al. (2012) is the first who achieved object recognition algorithm (AlexNet) that learns features directly from scratch (dataset). Other studies succeed based on transfer learning. For example, Trier et al. (2017) presents a novel approach based on AlexNet to estimate paper format (pixel per inch) and resolution of digitized document image. Akilan et al. (2017) presents a fusion between AlexNet, VGG-16 and Inception-V3 networks to improve image classification in comparison of single network. Pauly et al. (2017) focuses research on cracks by choosing AlexNet and notice that more network is deep more accuracy is higher, and accuracy depends on images location variance. Gopalakrishnan et al. (2017) studies crack detection in Hot-Mix Asphalt and Portland Cement Concrete with VGG-16 that gives high performance.

2.1 Transfer Learning Using Pre-trained AlexNet Neural Network

Alexnet DCNN architecture (Krizhevsky et al. 2012) is composed of 25 layers: eight layers with learnable weights such as five are convolutional layers and three are fully connected layers (Pauly et al. 2017). DCNN can be implemented (Vapnik and Izmailov 2017) for transfer learning with high performance in several domains. In this paper, pavement vision classification is established in Matlab. It requires Neural Network Toolbox, Statistics and Machine Learning Toolbox, Computer Vision System Toolbox and Neural Network Toolbox Model for AlexNet. Thereby, automatic image recognition classify, in the first case, three classes: whether road pavement contains cracks, tearing, or neither of the two degradations. It needs to modify fully connected layers (fc) by three neurons instead of 1000 neurons) in step of Fig. 1. As water seeps through degradations (cracks and pullouts), the roadway impermeability is defective. For saving time classifying, a second classifier is trained to predict only between two classes (road pavement degraded or not degraded). Therefore, it needs to modify fully connected layers (fc) by two neurons instead of three neurons according to the Fig. 2.

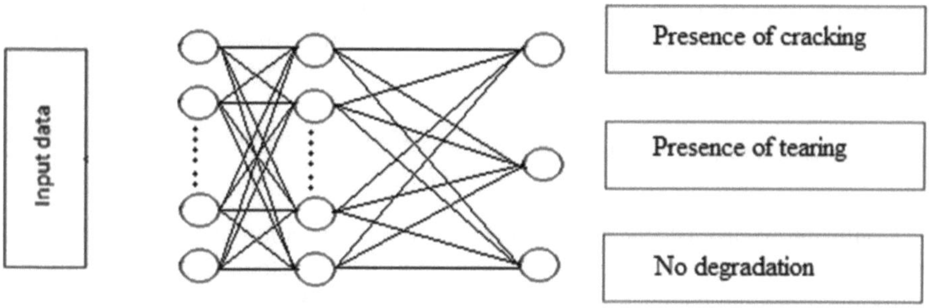

Fig. 1. Transfer learning using pre-trained AlexNet for classifying images into three classes

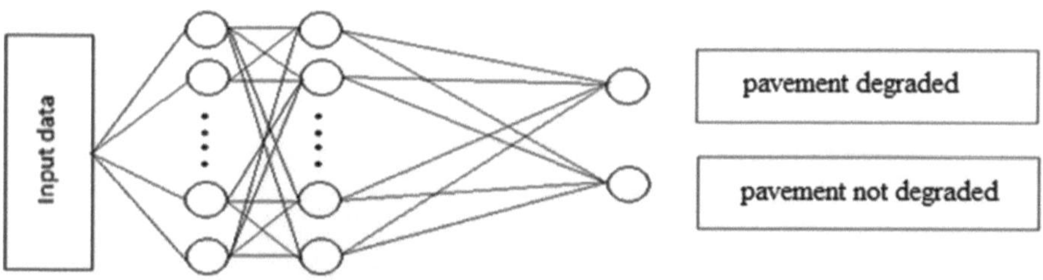

Fig. 2. Transfer learning for classifying roadway images into two classes

The image Data-store regroups two subfolders; each one contains RGB images with different sizes. The subfolders are named with labels referring to the classes (road pavement contains cracks or tearing or neither of the two degradations) in the first case and subfolders labels are referred to two classes (pavement degraded or pavement not degraded) in the second. To train network from AlexNet [1], images need to be pre-processed (resized and cropped) to match the expected input in (1) such as:

$$[\text{Width Height Samples\_Per\_Pixel}] = [227\ 227\ 3] \quad (1)$$

Once preprocessed, image Data-store is splitted into images for training (60%) and images for testing (40%) in order to evaluate network performance.

For training, algorithm-training option (Pauly et al. 2017) is useful for monitoring learning progress. In addition, it shows how fast and how accurate the training is done from launching training until getting 100% accuracy. Commonly, "stochastic gradient descent with momentum sgdm" algorithm option is set by default for training.

3 Results and Discussion

Training and testing are done using Intel (R) Core (TM) i5-7200U CPU @ 2.50 GHz 2.70 GHz with 400 GB. It processed the first case in 5510 s and the second in 1001 s only. After network training, images testing are classified using trained network to establish several verification for its performance and accuracy results.

Classification Accuracy is the ratio of total correct predictions to total predictions in testing images (Krizhevsky et al. 2012; Pauly et al. 2017). In the first case, accuracy reaches 70% according to Eq. (2).

$$Network\ Accuracy\,(case\ 1) = \frac{Total\ of\ Correct\ Predictions}{Total\ of\ predictions} = \frac{81}{114} = 71\% \quad (2)$$

Confusion Matrix (Pauly et al. 2017) is a matrix for confrontation of output and differentiates all cases between predictions and actual output. Thereby, confusion matrix describes the complete performance of the network (Fig. 3).

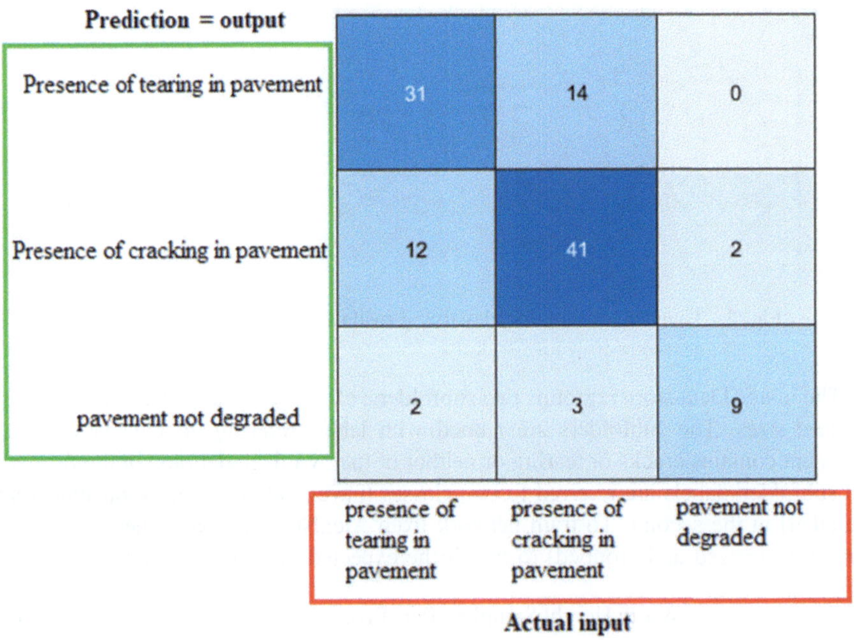

Fig. 3. Confusion matrix for confrontation between three classes

Transfer learning in the first case shows its efficiency. Figure 4 illustrates the gap between the input (images labels) and the output of the network (network predictions). According to the confusion matrix (Fig. 3), the gap is occurred only between two classes (presence of cracking and presence of tearing). It proves that there is a correlation between the two classes (Fig. 5) and it is true because cracking can be developed and become tearing. Consequently, the network could not differentiate between a crack with large opening and a tearing at the beginning of its evolution (Fig. 6). As the classes (cracking and tearing) are correlated, the second classifier shrink on two classes (pavement degraded and pavement not degraded) which represents the second case.

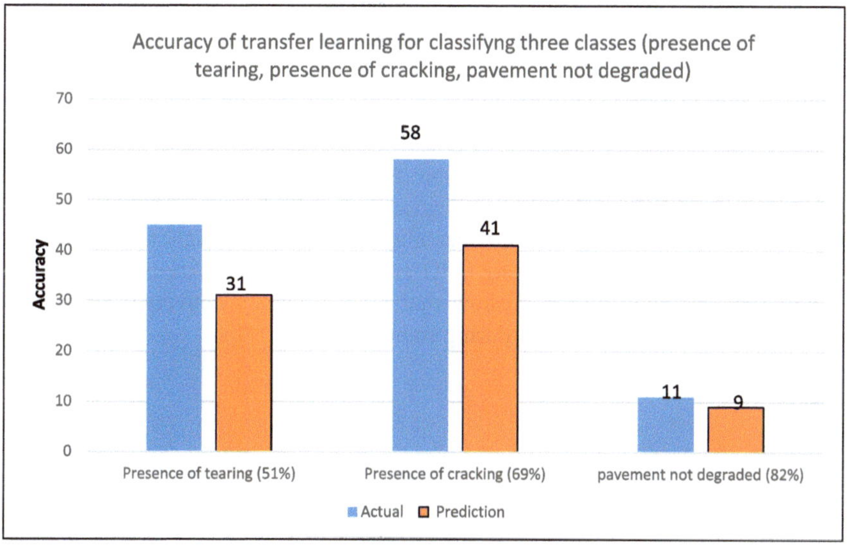

Fig. 4. Accuracy for the transfer learning network in case 1 (three classes)

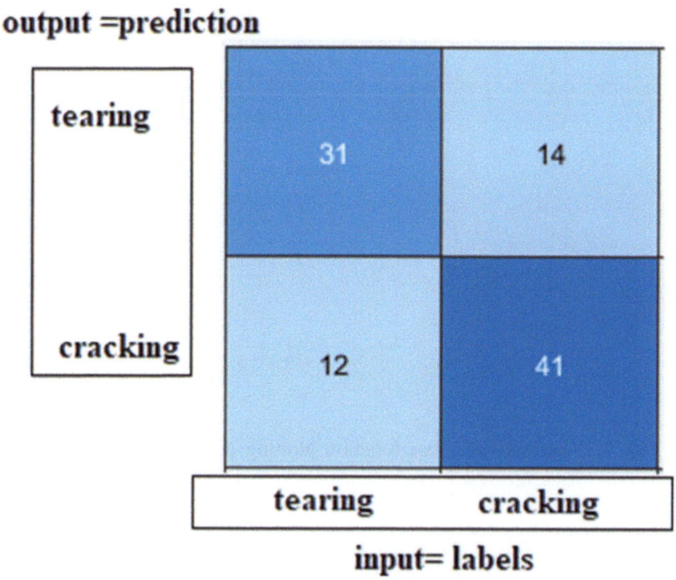

Fig. 5. Confusion matrix of tearing and cracking between actual labels and network predictions

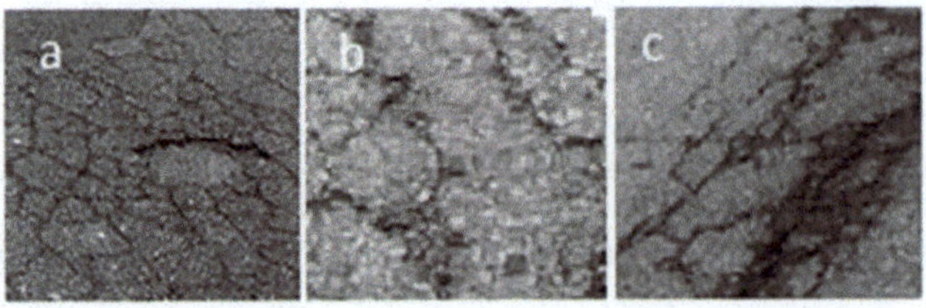

Fig. 6. Example of misunderstanding images between tearing and cracking (a. cracks opening grows and local tearing appears in wearing course, b. tearing between close cracks, c. branching of mesh cracks with a progressive tearing)

The second case is binary classification; actual and predicted output are classified into two classes: YES (road pavement degraded = risk of permeability) or NO (road pavement not degraded = good impermeability). Training on single CPU lasts 1001 s but it may take less time while using GPU graphics processor. Accuracy plotting and loss function illustrated in Fig. 7 show that more accuracy converge the more loss decreases until reaching zero loss during training.

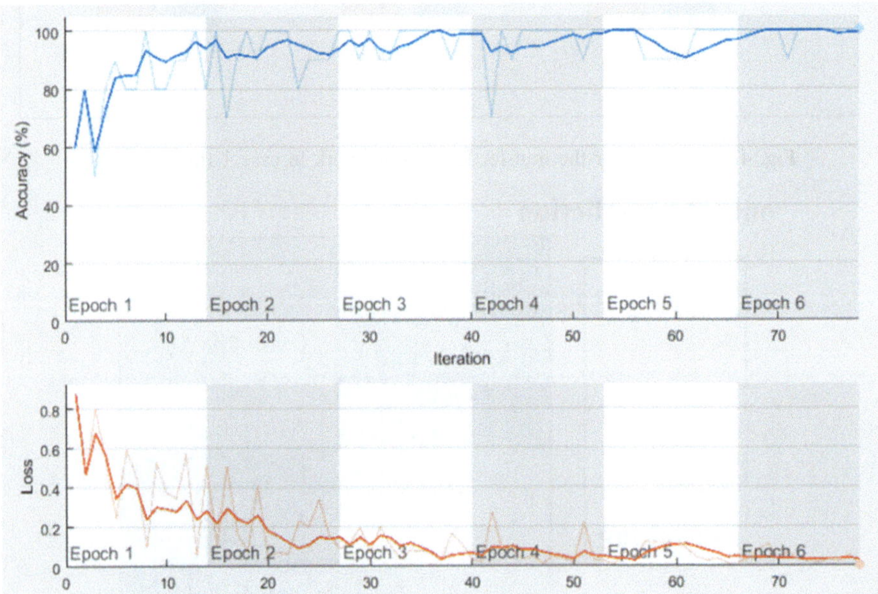

Fig. 7. Accuracy and loss function plotting during network training

Confusion matrix is given in Fig. 8 on testing the trained network model of 92 testing images.

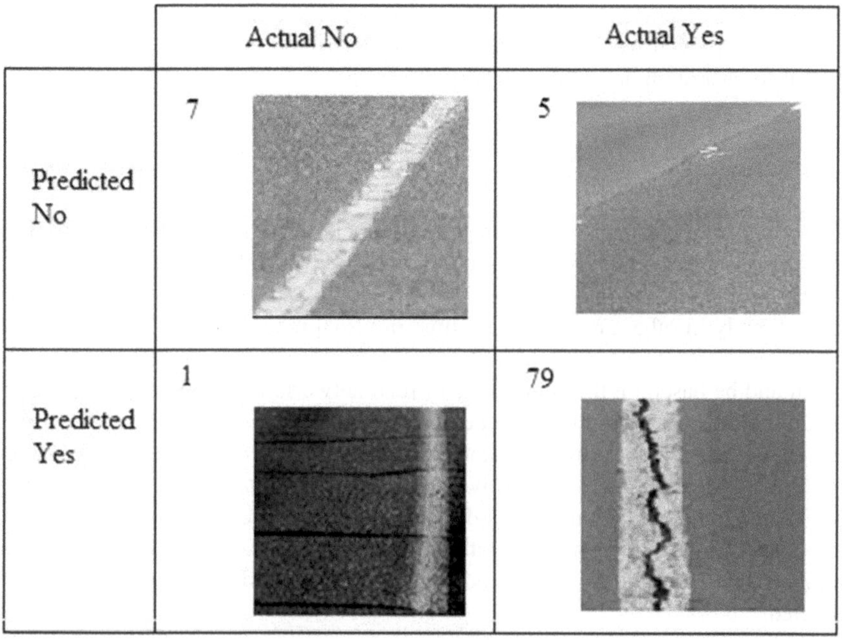

Fig. 8. Confusion Matrix for classifying two classes (pavement not degraded No and pavement degraded Yes)

The results of predictions in comparison with actual output can be divided to 4 cases (Pauly et al. 2017 and Imagenet):

- True Positives: The cases where prediction is YES and the actual output is YES. **True positive = 79.**
- True Negatives: The cases where prediction is NO and the actual output is YES. **True Negatives = 5**
- False Positives: The cases where prediction is YES and the actual output is NO. **False Positives = 1**
- False Negatives: The cases where prediction is NO and the actual output is NO. **False Negatives = 7**

According to the second classifier, convolutional neural network gives accuracy of 94% (high accuracy network) according to Eq. (3)

$$Accuracy, case\,2 = \frac{TP + FP}{TP + TN + FP + FN} = 94\% \quad (3)$$

It classifies all cases distinguishing between absence of infiltration risk and presence of risk. This automatic recognition affects directly road users' safety and the human and material damage level specific to the neighboring population or even the socio-economic level of the region. As a result, this automatic recognition has a direct impact on the collection of risk areas, especially the road network, in order to activate proposition of preventive solutions to block the deterioration of road conditions and to protect it to ensure an acceptable quality of service.

4 Conclusion

In this paper, two classifiers are trained to automatic recognition of pavement degradations depending on the purpose (summarily if pavement is degraded or not and precisely to distinguish between cracks and pullouts and good pavement). It can be applied directly in other areas and earn time in visual record instead of human vision. For sustainable road network and high users safety, program maintenance should be frequent and be based on the actual state of road network. In addition, before and after the rainy periods, roadway drainage network needs maintenance because water is the predominant triggering factor of land instability. Efficiency of drainage systems upstream and downstream of roadways is crucial in addition to the reparation of the downstream abutment degraded by undermining and superficial erosion of thalwegs. If specific land movements is noticed without pavement experiencing degradation due to infiltration of water across the roadway, a particular detailed study should be conducted.

Acknowledgments. The authors thank the anonymous readers for their comments and suggestions that have helped to improve the quality of manuscript.

References

Maurer, G.: L'érosion dans le Rif et le Prérif. Revue de géographie du Maroc **6**, 87–116 (1964)

Chalouan, A.: Les nappes ghomarides (Rif septentrional Maroc). Un terrain varisque dans la chaîne alpine. Ph.D. thesis, Univ. Louis Pasteur, Strasbourg (1986)

El Fellah, B.: Eboulement rocheux dans le Paléozoïque du Rif: Présentation d'un cas sur la route Oued Lao-Jebha. 7Eme Congrès International de l'AIGI, Lisboa, Portugal, pp. 3927–3931 (1994)

Fares, A.: Essai méthodologique de la cartographie des risques naturels liés aux mouvements de terrain. Application à l'aménagement de la ville de Taounate (Rif, Maroc). Thèse de Doctorat de l'Université de Franche Comté, 177 p. (1994)

El Khattabi, J.: Démarche méthodologique pluridisciplinaire intégrant une approche transversale pour l'étude des instabilités des versants: application aux versants du Rif Central (Maroc). Thèse de Doctorat, Université d'Artois, 258 p. (2001)

El Khattabi, J., Carlier, E.: Tectonic and hydrodynamic control of landslides in the northern area of the Central Rif, Morocco. Eng. Geol. **71**(3), 255–264 (2004)

Sossey Alaoui, F.: Traitement et intégration des données satellitaires optiques et radar dans un SIG en vue de l'obtention de carte de l'aléa liéaux instabilités de terrain dans la péninsule de Tanger (Rif Septentrional, Maroc). Thèse de Doctorat, Université Mohamed V, Faculté des Sciences, Rabat (2005)

Benmakhlouf, M., Galindo-Zaldívar, J., Chalouan, A., Sanz de Galdeano, C., Fedal Ahmamou, M., López- Garrido, A.C.: Inversion of transfer faults: the Jebha-Chrafate fault (Rif, Morocco). J. Afr. Earth. Sci. **73–74**, 33–43 (2012)

Ouadif, L., Bahi, L., Baba, K., Akhssas, A.: Study of stability of a highway fill application of the analytical and the finite elements methods. Int. J. Eng. Res. Appl. **2**(1), 904–910 (2012)

Baba, K., Bahi, L., Ouadif, L., Akhssas, A.: Slope stability evaluations by limit equilibrium and finite element methods applied to a railway in the Moroccan Rif. Open J. Civ. Eng. **2**, 27–32 (2012)

Mastere, M.: La susceptibilité aux mouvements de terrain dans la province de Chefchaouen: analyse spatiale, modélisation probabiliste multi-échelle et impact sur l'aménagement et l'urbanisme. Thèse de Doctorat, Université de Bretagne Occidentale (2011)

Mastere, M., Lanoë, B., Ait Brahim, B., El Moulat, M.: A linear indexing approach to mass movements susceptibility mapping. Revue Internationale de géomatique **25**(2), 245–265 (2015)

El Fahchouch, A.N., Brahim, L.A., Raji, O., Khouakhi, A.: Apport du SIG et de la télédétection dans la modélisation spatiale de la susceptibilité aux mouvements de terrain dans la région d'Al Hoceima, Rif Oriental, Maroc. Afrique Sci. **11**(2), 44–57 (2015)

Boujaaj, A., Bahi, L., Ouadif, L., Baba, K.: A methodology based on GIS for 3D slope stability analysis. Int. J. Eng. Technol. **8**(5), 2259–2264 (2016)

El-Fengour, A.: Landslide susceptibility assessment in the Amzaz Valley, Central Rif, Morocco. Doctorate thesis in Porto University (2016)

Zou, Q., Cao, Y., Li, Q., Maob, Q., Wangc, S.: CrackTree: automatic crack detection from pavement images. Pattern Recognition Letters **33**(2012), 227–238 (2012)

Saranya-vijaya, Geethalakshmi, S.N.: A survey on crack detection using image processing techniques and deep learning algorithms. Int. J. Pure Appl. Math. **118**(8), 215–220 (2018)

Laurent, J., Hébert, J.F., Talbot, M.: Automated detection of sealed cracks using 2d and 3d road surface data. Pavemetrics Systems Inc., Canada (2010)

Krizhevsky, A., Sutskever, I., Hinton, G.: Imagenet classification with deep convolutional neural networks. In: Advances in Neural Information Processing Systems, pp. 1097–1105 (AlexNet) (2012)

Simonyan, K., Zisserman, A.: Very deep convolutional networks for large-scale image recognition. In: International Conference on Learning Representations, vol. abs/1409.1556. (VGG-16 and VGG-19) (2014)

Deng, L., Hinton, G., Kingsbury, B.: New types of deep neural network learning for speech recognition and related applications: an overview. In: IEEE International Conference on Acoustics Speech and Signal Processing (ICASSP). (ResNet) (2013)

Szegedy, C., Liu, W., Jia, Y., Sermanet, P., Reed, S.E., Anguelov, D., Erhan, D., Vanhoucke, V., Rabinovich, A.: Going deeper with convolutions. CoRR, vol. abs/1409.4842 (2014)

Trier, F., Afzal, M.Z., Ebbecke, M., Liwicki, M.: Deep convolutional neural networks for image resolution detection. In: Proceedings of 4th International Workshop on Historical Document Imaging and Processing, Kyoto, Japan, 10–11 November 2017 (HIP2017) (2017)

Pauly, L., Hogg, D., Fuentes, R.: Deeper networks for pavement crack detection. In: Proceedings of the 34th ISARC, 34th International Symposium in Automation and Robotics in Construction, Taipei, Taiwan. IAARC, pp. 479–485 (2017)

Akilan, T., Wu, Q.M.J., Yang, Y., Safaei, A.: Fusion of transfer learning features and its application in image classification. In: IEEE 30th Canadian Conference on Electrical and Computer Engineering (CCECE) (2017)

Gopalakrishnan, K., Khaitan, S.K., Choudhary, A., Agrawal, A.: Deep convolutional neural networks with transfer learning for computer vision-based data-driven pavement distress detection. Constr. Build. Mater. **157**(2017), 322–330 (2017)

Vapnik, V., Izmailov, R.: Knowledge transfer in SVM and neural networks. Ann. Math. Artif. Intell. **81**, 3 (2017). https://doi.org/10.1007/s10472-017-9538-x

Matlab https://www.mathworks.com/

ImageNet. http://www.image-net.org

Asphalt Concrete Performance Combining Use of Increased Reclaimed Asphalt Content and Warm Mix Additives

Pavla Vacková[1], Jan Valentin[1(✉)], and Meor O. Hamzah[2]

[1] Faculty of Civil Engineering, Department of Road Structures, CTU in Prague, Thakurova 7, 166 29 Prague, Czech Republic
{pavla.vackova.1, jan.valentin}@fsv.cvut.cz
[2] The School of Civil Engineering, Universiti Sains Malaysia, Pulau Pinang, Malaysia
cemeor@yahoo.com

Abstract. In developed countries there is a clear shift from new pavement structures to rehabilitation works which usually use the mill and fill method by replacing some of the asphalt layers. This process on one hand creates material which has to be milled and on the other hand requires new material for producing asphalt mixtures which are used back to the pavement structure. To improve the economic efficiency and reduce environmental impacts reuse or recycling are presently supported in elevated extent. To make the highest possible use of reclaimed asphalt rejuvenation is seen as one of the options. On the other hand excessive heating of reclaimed asphalt should be avoided to not overburn the bitumen or even to avoid additional ageing of already aged material. Combining effects of rejuvenation and warm mix technologies seems as an interesting option. In this respect a trial section was designed including 13 different options of warm mix asphalt with 0–50% reclaimed asphalt. The asphalt mixtures differed in amount of used reclaimed asphalt (RA) 0/11 mm (0%, 30% and 50%), the gradation of used paving grade bitumen (50/70,70/100) and warm mix asphalt additives or used techniques. In asphalt mixtures there were used bituminous binders with selected low temperature additive, which have primarily the function of reducing the production temperature, or the benefit in terms of improving the properties of the asphalt mixture, the adhesion improvement (in some cases) and in at least one of the cases the potential for reducing the characteristic odor of asphalt mixtures. Additives based on organosilanes and different types of synthetic waxes (FT paraffin, amide wax, PE wax) have been used. Additionally foamed bitumen technology was applied as well. The asphalt mix performance was assessed by term of stiffness, resistance to permanent deformation, water susceptibility and thermal induced as well as fatigue cracking. The key findings will be presented by the paper.

1 Introduction

One of the most challenging current topics is sustainable development. The mankind product enormous amount of waste and greenhouse emissions. There is an important influence from authorities to reduce both of these "by-products". A need to reduce

S. El-Badawy and R. Abd El-Hakim (Eds.): GeoMEast 2018, SUCI, pp. 145–157, 2019.
https://doi.org/10.1007/978-3-030-01908-2_12

global greenhouse emissions is in order to reduce climate changes and a need to reduce waste disposals is in order to reduce the land occupation (NAPA 2012).

At the same time, there is need for ongoing development of progressive technologies and materials as a result of increased traffic volumes and often extreme climatic conditions in order to ensure a sustainable, environmentally and financially efficient road network (EAPA 2014).

Warm mix asphalt technology appears to be a promising solution from point of view of sustainability. Common hot asphalt mixtures, which covers in the Czech Republic as in many other European countries about 80–90% of all road surface layers (White book 2011), are manufactured and paved in range of 150–180 °C, depending on used bituminous binder (Sdružení pro výstavbu silnic 2018). High temperatures which are necessary in manufacturing and compacting of hot asphalt mixtures have a negative impact on the environment and on construction workers, who are almost daily in contact with these materials (Rühl and Musanke 2002).

From these reasons there are many research studies and construction companies working and testing alternative technologies and manufacturing processes targeting to reduce the energy consumption and greenhouse emissions while the asphalt mixture properties are remained same or improved (e.g. Capayova et al. 2017, Cardone et al. 2009, Hill 2011, Zhao et al. 2013).

The next step towards energy consumption and greenhouse emissions reduction is a combination of use of reclaimed asphalt (RA) and warm mix additives. The advantage of this combination is a lowering of heating temperature of RA (Hamzah et al. 2010). Reclaimed asphalt contains aged bituminous binder, which is oxidized and hardened. Heating to elevated temperatures causes additional ageing and hardening (Cavalli et al. 2018). Other advantage of warm mix additives is improving of workability. Additionally, some of the warm mix additive are able to "modified"/soften the aged binder in RAP and therefore to rejuvenate it (Benešová et al. 2017).

2 Assessed Variants of Warm Mix Asphalts

Within the presented trial section the asphalt plant produced 13 different warm mix asphalts of type AC_{surf} 11+. The trial section was set to verify different technical solutions and possibilities for warm mix asphalts designed with high content of reclaimed asphalt.

The assessed asphalt mixtures differed in used content of reclaimed asphalt (RA) 0/11 mm (0%, 30% and 50%), in the applied paving grade bitumen (50/70 or 70/100) and in used warm mix additives. Additionally there were variants using the foamed bitumen technique. The primarily function of used additives was the possibility to reduce the mixing temperature. Nevertheless additional benefits were expected in terms of improving some of the asphalt mix properties. Additives based on organosilanes (ZT) or different types of synthetic waxes (SR, DE, IN, and PS) were selected and used. Just for clarification the used product abbreviations are used to avoid commercial trade names or trademarks. For this reason full names are not mentioned. In some variants foamed bitumen technology was used instead of regular bituminous binder and the mentioned additives. The foam bitumen technology enables the reduction of mixing temperature by creating a low-viscous bitumen foam, which has large free surface and very low viscosity at given temperature of production and paving of an asphalt mixture (Ruckel et al. 1983).

In the case of all asphalt mixtures containing reclaimed asphalt a REOB (recycled engine oil bottoms) based rejuvenators have been used as well. The function of these additives is to partial restore the properties of aged bitumen in the used reclaimed asphalt, i.e. rejuvenation of aged and degraded bitumen. In most of the asphalt mixtures containing reclaimed asphalt SX rejuvenator was used. The exception were asphalt mixtures, where combined additive ST (#6) or SE (#7) was used. In these two cases the rejuvenator SX was replaced and in the case of variant #7 no additional warm mix additive was used. It has to be stressed that the effect of the rejuvenators on bituminous binders was not experimentally analyzed. The used content of rejuvenator was based on previous experience of the mixing plant and therefore the paper does not contain any additional information related to rejuvenator content optimization process.

The summary of all AC_{surf} 11+ asphalt mix variants including the specification of used warm mix additives and input paving grade bitumen is given in Table 1. In this experimental study assessment for selected asphalt properties for each asphalt mixture was for better and quicker understanding further marked by an identification code using #1 up to #13 according to placement through the trial section. This identification is used by all further presented graphs and tables.

Table 1. Overview of assessed ACO 11+ mixture

Mix ID	Bitumen	RA content	Additive	Additive based on
#1	70/100	0%	Foamed bitumen (+SX)	
#10	50/70	30%	Foamed bitumen (+SX)	
#2	70/100	30%	Foamed bitumen (+SX)	
#4	50/70	50%	SR (+SX)	Wax
#5	50/70	50%	SR (in the bitumen) (+SX)	
#6	50/70	50%	ST	
#7	50/70	50%	SE	Rejuvenator
#9	50/70	50%	PS (+SX)	Wax
#11	50/70	50%	Foamed bitumen (+SX)	
#13	50/70	50%	ZT (+SX)	Silanes
#3	70/100	50%	Foamed bitumen (+SX)	
#8	70/100	50%	IN (+SX)	Wax
#12	70/100	50%	DE (+SX)	Wax

The asphalt mixtures were for better understanding and easier comparison further ordered in smaller logical groups. The grouping was based on the similarities of input materials or following the impact of used technology for warm mix asphalts.

3 The Range of Performed Testing

Asphalt mix AC_{surf} 11+ variants were always designed with an identical grading curve and same bitumen content, i.e. including the conversion of the virgin binder depending on the quantity of reclaimed asphalt used. It is also important to accentuate that the mixtures were produced by the same mixing plant and that in the case of using reclaimed asphalt in the mixture same source of this material was used, including same process of material customization (screening and homogenization). Asphalt mixtures used for further presented experimental results were taken gradually during the construction progress and paving of each variant on the trial section.

The following characteristics were determined for all variants of AC_{surf} 11+ asphalt mixture, including subsequent test data evaluation

- **Volumetric characteristics** (maximum bulk densities according to the European standard *EN 12697-5*, bulk densities according to the European standard *EN 12697-6*, voids contents of asphalt mixtures according to the European standard *EN 12697-8*);
- **Resistance of asphalt mixture to water immersion** according to the European standard *EN 12697-12*. This test procedure was extended by a modified procedure according to the U.S. method given in the standard *AASHTO T 283*;
- **Stiffness modulus determined** according to the European standard *EN 12697-26* by non-destructive test method (IT-CY) for selected test temperatures. According to the practice used in the Czech Republic usually 0 °C for cold seasons, 15 °C for average spring or autumn seasons and 27 °C as average summer temperature are used.
- **Resistance to thermal crack propagation** according to the procedure given in the European standard *EN 12697-44:2011* (three-point bending test performed on semicircular test specimens done usually at the temperature of 0 °C). Compared to the given technical standard test specimens with a diameter of 100 mm were used compacted either according to the European standard *EN 12697-30*;
- **Determination of resistance to permanent deformations** according to the European standard *EN 12697-22+A1* performed on small test device in an air bath with a temperature of 50 °C.

4 Results and Discussion

4.1 Volumetric Characteristics

The results of volumetric characteristics are summarized in Table 2 and in Fig. 1. The Fig. 1 also shows the limits for air voids content in case of AC_{surf} 11+ mixture in compliance with the requirements of the national specificities under *EN 13108-1*. The test specimens were compacted according to *EN 12697-33* with 2 × 50 blows by impact compactor at temperature of 130 °C.

From the point of view of maximum density, the results reflect the heterogeneous effect of reclaimed asphalt comprising aggregate different from that in the reference

Table 2. Volumetric characteristics of assessed AC_{surf} 11+ variants

		Maximum density (g cm^{-3})	(MT)	
			Bulk density (g cm^{-3})	Voids content (%)
#1	Foamed bitumen	2,607	2,483	4,8%
#10	Foamed bitumen	2,547	2,419	5,0%
#2	Foamed bitumen	2,544	2,459	3,3%
#4	SR	2,556	2,454	4,0%
#5	SR (in the bitumen)	2,545	2,433	4,4%
#6	ST	2,531	2,428	4,1%
#7	SE	2,527	2,392	5,3%
#9	PS	2,530	2,442	3,5%
#11	Foamed bitumen	2,592	2,430	6,2%
#13	ZT	2,578	2,435	5,6%
#3	Foamed bitumen	2,541	2,452	3,5%
#8	IN	2,549	2,443	4,2%
#12	DE	2,522	2,427	3,8%

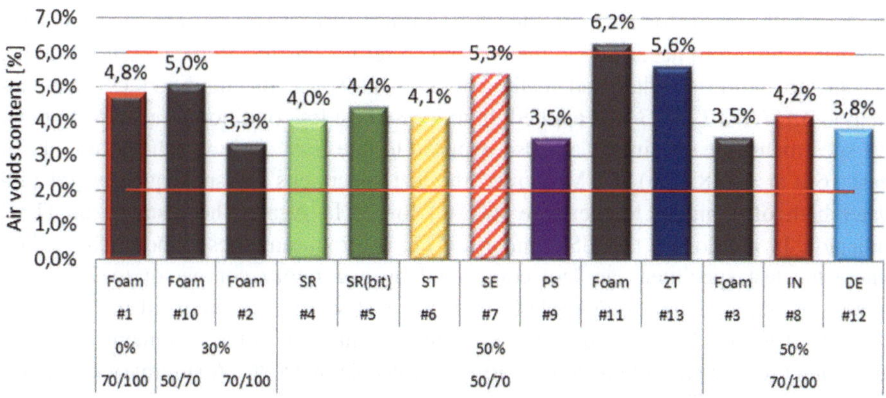

Fig. 1. Voids content results of assessed AC_{surf} 11+ mix variants

asphalt mixture, which logically results in a different value in most cases. Because the reclaimed asphalt was not taken from a particular construction site but rather from a stockpile with material from different sites, the variability of the aggregates in the reclaimed asphalt can be quite large.

Air voids content of only one mixture variant (ACO #11–50% RA with foam bitumen) exceed the standard limit. To a certain point, the above might serve as an indicator of inappropriateness of paving grade 50/70 for foamed bitumen to be used in combination with reclaimed asphalt as also mixture ACO #10 demonstrates a higher voids content relative to mixture ACO #2 (the only difference is the type of paving grade bitumen).

It should also be pointed out that mixtures ACO #7, #10 and #13 fall within the voids content range of 5–6%-vol. (higher values). In the first case, this could be explained by the presence of rubber in the additive; the second case has been described above while the result is rather surprising in the last case with an organo-silane additive. This fails to correlate fully with the corresponding foreign findings. We have been unable to identify the cause for rather average results in this case. The remaining mixtures fall within a relatively narrow voids content interval of 3.0–4.5%-vol.

4.2 Moisture Susceptibility

Another determining and commonly assessed characteristic currently monitored in asphalt mixtures and used as an indicator of durability is the moisture susceptibility (sensitivity to water immersion). The basic technical standard for the test is *EN 12697-12* which stipulates the moisture susceptibility ratio (ITSR). The indirect tensile strength of dry test specimens, i.e. specimens not exposed to the effect of water immersion, is usually listed along with the parameter.

Based on a number of earlier findings, indirect tensile strength of dry specimens should correlate well with the stiffness modulus values as presented in the next part of the paper. The test is performed with test specimens compacted by 2 × 25 impacts in the Marshall compactor at 15 °C and 50 mm/min loading speed. The selected temperature is preferred according to the preference set for this test in the Czech Republic.

On top of the standardised method in *EN 12697-12,* the modified US standard method *AASHTO T283-3* was applied. In this case besides conditioning water-saturated test specimens, the specimens are also exposed to one frost cycle which should take at least 16 h at − (18 ± 3) °C. Moreover, the test specimens are subsequently put in a water bath of a higher temperature (60 °C) for 24 h. Again, the result is an ITSR indicator. In relation to the US standard, CTU in Prague applies a modified method where the test specimens are not compacted by a gyratory compactor to 7% voids content as required by the AASHTO standard; test specimens prepared to the specifications of *EN 12697-12* are used instead. Subsequently, all of the specimens are tested for indirect tensile strength according to *EN 12697-23* at 15 °C. A summary of the test results is given in Table 3.

If 80% were chosen as the threshold value from the perspective of moisture susceptibility for EN 12697-12 test, then just mixtures #3 and #10 are reaching the limiting value. Although both of them are asphalt mixtures with foamed bitumen and various contents of reclaimed asphalt, the results provide no guidance concerning a tendency as the remaining variants with foamed bitumen application demonstrate good, or even very good level of water susceptibility. If moisture susceptibility was assessed according to the US test procedure, a number of variants confirm the generally observed trend of ITSR decrease in comparison to the European method. However,

Table 3. Indirect tensile strength results and the water susceptibility ratio

		Indirect tensile strength			ITSR	
		Dry (MPa)	EN (MPa)	AASHTO (MPa)	EN (%)	AASHTO (%)
#1	Foamed bitumen	2,20	1,80	1,73	81,9%	78,5%
#10	Foamed bitumen	2,18	1,76	1,55	80,9%	71,1%
#2	Foamed bitumen	1,76	1,68	1,95	95,3%	110,5%
#4	SR	1,90	1,86	1,71	97,8%	90,1%
#5	SR (in the bitumen)	1,97	1,88	2,49	95,5%	126,6%
#6	ST	1,25	1,33	1,45	106,4%	116,1%
#7	SE	1,59	1,46	1,42	91,9%	89,5%
#9	PS	1,99	1,93	1,79	97,4%	89,9%
#11	Foamed bitumen	1,91	1,74	1,69	90,9%	88,5%
#13	ZT	1,99	1,83	1,78	92,0%	89,8%
#3	Foamed bitumen	2,16	1,73	2,04	80,1%	94,2%
#8	IN	2,14	1,76	1,96	81,9%	91,3%
#12	DE	1,83	1,75	1,69	95,5%	92,4%

only two mixtures (#1 and #10) drop below 80% and in both cases, this is influenced by the lower value recorded already by the European test method as well. From the perspective of moisture susceptibility, it can be noted that the application of increased contents of reclaimed asphalt in combination with various additives lowering the working temperature has no negative effect on asphalt mixture durability. Table 3 offers even several values exceeding the 100% limit. Nevertheless, these results must be interpreted with caution. This is an occasional phenomenon which might be caused by the effects of water and temperature during test specimen conditioning as is not completely explainable, or by imprecisions in test temperature setting for the indirect strength test reading. A 1 °C difference in the test temperature, whether up or down, might influence the test values significantly. No cause could be determined in the case of the measurements taken and subsequent evaluation thereof.

4.3 Stiffness of Tested Asphalt Mixtures

The stiffness modulus is one of the fundamental deformation characteristics of asphalt mixture which is commonly taken into account in the flexible pavement design.

The stiffness modulus was determined non-destructively on cylindrical test specimens, of a 100 mm diameter, according to the specification of method C (IT-CY) as stipulated by *EN 12697-26*.

The stiffness modulus was measured at temperatures usually assessed in the Czech Republic (0, 15, 27 °C), with always at least 6 test specimens. The stiffness modulus values are summarised in Table 4 and Fig. 2.

Table 4. Stiffness modulus

		0 °C (MPa)	15 °C (MPa)	27 °C (MPa)	Thermal susceptibility
#1	Foamed bitumen	22,106	10,725	3276	6,7
#10	Foamed bitumen	20,921	8592	3223	6,5
#2	Foamed bitumen	17,764	8614	2430	7,3
#4	SR	22,505	8506	2966	7,6
#5	SR (in the bitumen)	19,487	7872	3006	6,5
#6	ST	15,648	5289	2076	7,5
#7	SE	16,987	7393	3225	5,3
#9	PS	20,617	10,113	4055	5,1
#11	Foamed bitumen	20,147	9132	2947	6,8
#13	ZT	19,515	9846	3085	6,3
#3	Foamed bitumen	19,429	8768	2898	6,7
#8	IN	19,268	9699	3442	5,6
#12	DE	20,249	10,111	3325	6,1

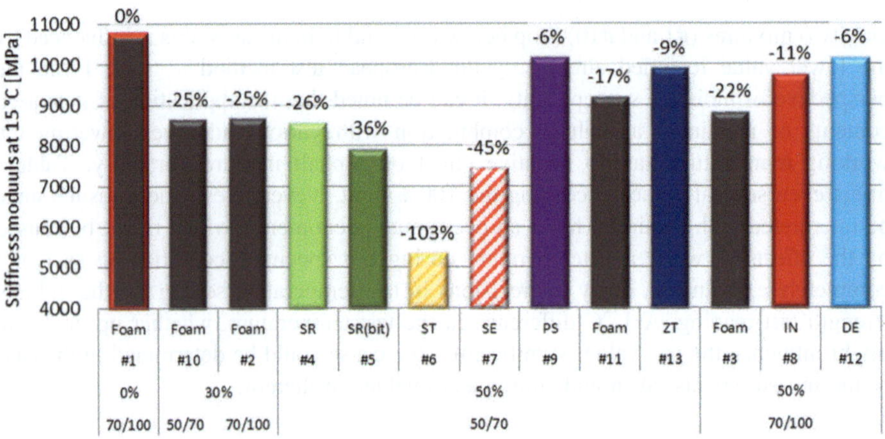

Fig. 2. Stiffness modules determined on assessed AC_{surf} 11+ mix variants

From the perspective of stiffness assessment at all selected test temperatures, the trend of all the individual variants at 0 °C is obviously similar to that at 15 °C with the only exception of mixture #9 (FT wax-based additive combined with some other chemicals). At the highest temperature measured, this applies to a certain degree as well… In case of thermal susceptibility, which is a ration of stiffness at the lowest and the highest temperature, the range extends from 5.1 (for mixture #9) to 7.6 (for mixture #4). The lower the value, the less susceptible to temperature changes the mixture is in terms of stiffness; logically, the mixture will resist thermal effects better. An interesting finding within the thermal range was the case of two variants with FT wax-based additives achieving differing thermal susceptibility scores. For the sake of comparison of the application method of SR, it should be pointed out that with respect to stiffness, the asphalt mixture where wax was added during asphalt mix preparation scored higher while the asphalt mixture with wax present directly in the bitumen had a lower thermal susceptibility value. It is still impossible to clearly state which method of application of the additive is more advantageous.

4.4 Resistance to Crack Propagation

One of the innovative methods incorporated in the harmonised European test standards is the determination of asphalt mixture resistance to crack propagation in semi-cylindrical test specimens, applying the three-point bending test. The test is carried out according to *EN 12697-44:2011*. The test specimens are modified compared to the standard. In this case the specimens compacted by impact compactor with 100 mm diameter were used. The specimens were cut on laboratory saw to reach the proper dimensions.

Essentially, the test consists of loading the test specimen evenly, at a constant speed (5 mm/min) up to the moment when a crack is formed in the weakened point within the specimen. The test uses the theory of fracture mechanics. To determine a weakened area in the compacted asphalt layer simulated in the laboratory by a suitable test specimen, a thin, 10 mm deep notch is made in the centre of the base of the test specimen. A load is applied against the notch to the point of failure (crack propagation) of the test specimen, or, according to the requirements of the European standard, just the moment of reaching the maximum force when a crack appears in the specimen is monitored (crack initiation). By recording the maximum force and deformation up to the point of test specimen failure the data are used for the calculation of the maximum stress/strain and fracture toughness (critical value for fracture to occur).

The test was always performed with at least 6 semi-cylindrical test specimens for which stiffness modulus had been determined. The test specimens prepared were conditioned for at least 8 h at the test temperature of 0 °C.

The results are summarized in Fig. 3

Figure 3 clearly shows the differences in the behaviour between the assessed asphalt mixtures. The lowest fracture toughness was demonstrated by the mixture with foamed bitumen made from paving grade 50/70 and 30% reclaimed asphalt which had fared poorly in other tests, too. Contrastingly, by far the best result with respect to fracture toughness was achieved by mixture #7 containing crumb rubber based additive. To a point, the result was to be expected.

4.5 Resistance to Permanent Deformations (Rutting)

The characteristics of resistance to permanent deformations of assessed AC_{surf} 11+ mix variants (wheel tracking test) were determined according to *EN 12697-22* using a small test device with air bath and a test temperature of 50 °C. The tracked and derived test parameters are the maximum proportional rut depth (PRD_{AIR}) and maximal wheel-tracking slope (WTS_{AIR}). In parallel the mean rut depth after 5000 and 10,000 cycles was recorded as well. The test was done on slabs with standardized dimension of 26 × 32 × 4 cm. The results are summarized in Table 5 and by Fig. 4. In the case of

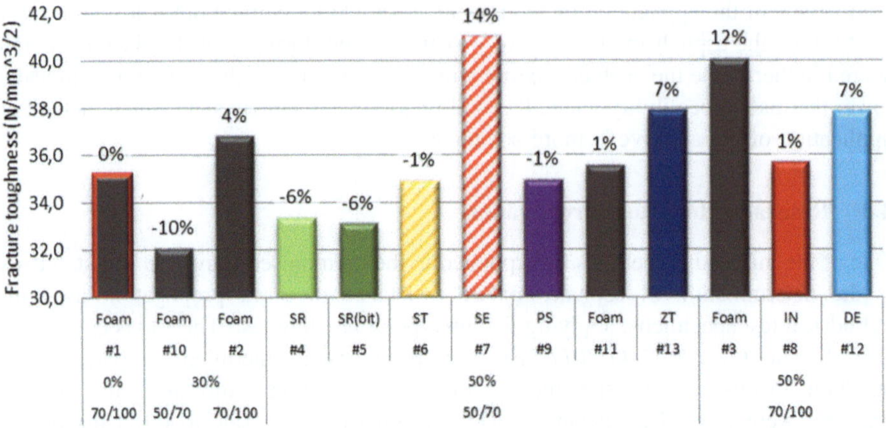

Fig. 3. Resistance to crack propagation of AC_{surf} 11+ mix variants at T = 0 °C

asphalt concrete type AC_{surf} 11+ the national appendix to product standard EN 13108-1 does not specify requirements on threshold values for parameters WTS_{AIR} a PRD_{AIR}. It is only required to record the results. For comparability purposes of all tested asphalt mix variants it is possible to use parameters which are by the same standard required for AC_{surf} 11S mixture (maximum value for WTS_{AIR} = 0.07 mm/10^3 cycles and maximum value for PRD_{AIR} = 5.0%).

It is obvious from the test results, that the best resistance to permanent deformation is reach by variants #4 (SR added during the asphalt mix production), #7 (SE) and #8 (IN), whereas the variant with SE additive containing crumb rubber achieves the lowest value in case of both parameters. It is necessary to emphasize, that in case of using this additive the presence of rubber is important with respect to strain behavior. The rubber acts in certain extent as a modifier and therefore differentiates from all other tested variants where some additives based on synthetic waxes can have stiffening effect, but they will primarily stay additives dedicated to reducing the production and paving temperature of an asphalt mix. Interesting is again the comparison of dosing effect in case of SR. In this case the use of bitumen doped by this additive has slightly better results for all considered parameters.

The worst results have been shown for variants #2, #3 a #10. For the first two mixtures the explanation can be given by the combination of used reclaimed asphalt and foamed bitumen based on paving grade 70/100, the third case bitumen 50/70 is used but the combination with 30% reclaimed asphalt for some reason does not seem to be an optimum solution. It should be additionally stated that from the perspective of thermal crack propagation or fatigue behavior the use of a softer bitumen for foamed bitumen technology must not be necessarily detrimental, rather conversely. Therefore it is important not to focus only on one material characteristic, but always to understand the behavior of an asphalt layer as a complex issue of different performance based effects.

Table 5. Rutting test characteristics for assessed AC_{surf} 11+ mix variants

		Mean rut depth after		WTSAIR	PRDAIR
		5000 cycles (mm)	10000 cycles (mm)	(mm)	(%)
#1	Foamed bitumen	1,19	1,34	0,029	3,0%
#10	Foamed bitumen	1,38	1,61	0,047	3,4%
#2	Foamed bitumen	1,40	1,66	0,052	3,5%
#4	SR	1,16	1,29	0,026	2,9%
#5	SR (in the bitumen)	1,44	1,61	0,033	3,6%
#6	ST	1,07	1,24	0,034	2,6%
#7	SE	0,86	0,98	0,025	2,2%
#9	PS	1,77	1,91	0,029	4,5%
#11	Foamed bitumen	1,06	1,21	0,029	2,7%
#13	ZT	1,59	1,74	0,030	4,0%
#3	Foamed bitumen	1,48	1,73	0,051	3,7%
#8	IN	1,24	1,37	0,027	3,1%
#12	DE	1,26	1,44	0,036	3,1%

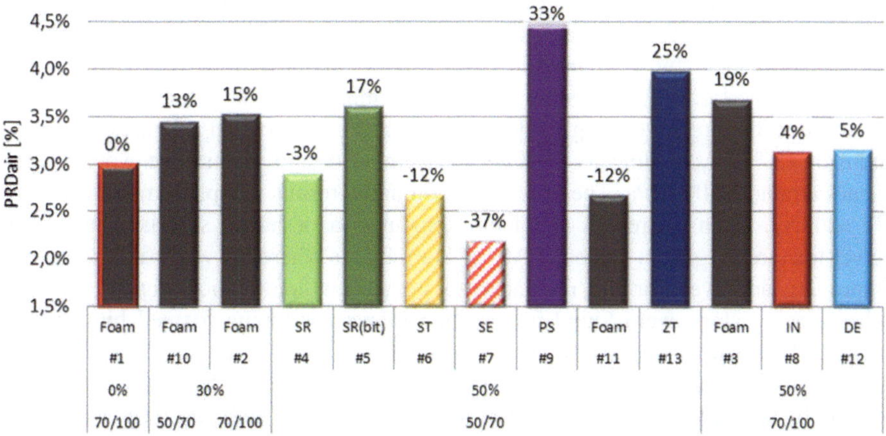

Fig. 4. Rutting test characteristics for assessed ACsurf 11+ mix variants

In general it can be stated, that all tested asphalt mixtures show even in case of using higher reclaimed asphalt content very good resistance to permanent deformations.

5 Summary

The realized laboratory comparison of extended number of warm mix asphalt variants combined with higher content of heated reclaimed asphalt delivers several valuable findings. Besides the verification that an asphalt mixture can be mixed and compacted at temperatures which by arrival of fresh mix at the construction site have a temperature of 130 °C with outside temperatures in a range of 10–17 °C, the key benefits are especially in the results with several performance based characteristics. In this text some of them have been presented in extended range.

Focusing firstly on volumetric characteristics, all assessed mixtures fulfilled required threshold values. Just the limit was only reached by variants with foamed bitumen 50/70 combined with 50% reclaimed asphalt. Obvious and repeatedly proven is the fact about impact caused by the selected laboratory compaction method.

Second fundamental characteristic is represented by the resistance of asphalt mixture to water immersion which is used for expressing asphalt mix durability. Also in this case there is no evidence about a more significant difference between the variants. All tested variants would fulfill a threshold value of 80%. A more outstanding worsening (moisture susceptibility) was not found even in the case of testing the mix variants according to the modified U.S. approach for water immersion including one frost cycle. The results do not confirm sometimes raised doubts related to impaired adhesion quality between bitumen and aggregate particles if lowered mixing and compaction temperatures are used in case of WMA applications.

Another performed test was the stiffness determination at three test temperatures. It is surprised how relatively high values are reached by almost all mixtures at temperature of 15 °C. According to the Czech pavement design manual we would expect values around 7.500 MPa. The only exception in the results is asphalt mixture #6 with ST additive. From the results of fracture toughness as a certain surprise can be seen lower fracture toughness values of AC_{surf} 11+ variants where SR was used as viscosity improving additive. For the resistance to permanent deformation higher values of wheel-tracing slope were found in case of variants combining foamed bitumen and reclaimed asphalt. Nevertheless, this rule cannot be applied absolutely. For the second characteristic the situation was different and higher values of PRD_{AIR} were demonstrated by asphalt mixtures with two different chemical additives (FT wax based additive and organo-silane based additive).

Despite of that, all assessed variants fulfilled requirements, which are set for higher quality class of asphalt concrete mixture.

Acknowledgments. This paper was elaborated within the activities of the Centre for effective and sustainable transport infrastructure (CESTI)—TE01020168 as supported by the Technology Agency of Czech Republic.

References

Benešová, L., Mastný, J., Valentin, J.: Effect of new types of synthetic waxes designed for low-temperature asphalt mixtures. In: Bearing Capacity of Roads, Railways and Airfields. Taylor & Francis Group, London, pp. 335–342 (2017). ISBN 978-1-138-29595-7

Capayova, S., Zuzulova, A., Hodakova, D., Schlosser, T.: Influence of chemical additives on bitumen binder properties. In: Proceedings of the 17th SGEM 2017 International Multidisciplinary Scientific Geoconference, Conference Proceedings Vol. 17, Energy and Clean Technologies No. 41, STEF92 Technology Ltd., Sofia, Bulgaria, pp. 487–494 (2017). ISBN 978-619-7408-06-5

Cardone, F., Pannunzio, V., Virgili, A., Barbati, S.: An evaluation of use of synthetic waxes in warm mix asphalt. In: Proceedings of the 7th International RILEM Symposium ATCBM09 on Advanced Testing and Characterization of Bituminous Materials, Rhodes, Greece (2009)

Cavalli, M.C., Zaumanis, M., Mazzaa, E., Partla, M.N., Poulikakosa, L.D.: Effect of ageing on the mechanical and chemical properties of binder from RAP treated with bio-based rejuvenators. Compos. B **141**, 174–181 (2018)

European Asphalt Pavement Association: The use of Warm Mix Asphalt: EAPA—position paper. Brussels, Belgium (2014)

Hamzah, M.O., Jamshidi, A., Shahadan, S.: Effects of Sasobit® on the required heat energy and CO2 emission on blended asphalt binder incorporated with aged binder. Eur. J. Sci. Res. **42** (1), 16–24 (2010)

Hill, B.: Performance evaluation of warm mix asphalt mixtures incorporating reclaimed asphalt pavement. Research Report. University of Illinois at Urbana-Champaign (2011)

National Asphalt Pavement Association: Warm-Mix Asphalt: Best Practices, 3rd edn. (2012)

Ruckel, P.J., Acott, S.M., Bowering, R.H.: Foamed-asphalt paving mixtures: preparation of design mixes and treatment of test specimens. Transp. Res. Rec. **911**, 88–95 (1983)

Rühl, R., Musanke, U.: Bitumen Forum—Low-temperature asphalt, European Agency for Safety and Health at Work (2002)

Sdružení pro výstavbu silnic: Přehledy výroby a zpracování materiálů pro stavbu vozovek v roce 2017, Sdružení pro výstavbu silnic (2018). (in Czech)

White EU book on transportation: EU Office ČS, a. s. Businessinfo (2011). http://www.businessinfo.cz/cs/clanky/bila-kniha-eu-o-doprave-5164.html. Accessed 5 Sept 2017

Zhao, S., Huang, B., Shu, X., Woods, M.: Comparative evaluation of warm mix asphalt containing high percentages of reclaimed asphalt pavement. Constr. Build. Mater. **44**, 92–100 (2013)

Effect of High RA Content on Properties of Stone Mastic Asphalt (SMA) Mixtures – Trial Section

Pavla Vacková[1], Jan Valentin[1](✉), and Evangelos Manthos[2]

[1] Faculty of Civil Engineering, Department of Road Structures, CTU in Prague, Thakurova 7, 166 29 Prague, Czech Republic
{pavla.vackova.1, jan.valentin}@fsv.cvut.cz
[2] School of Civil Engineering, Aristotle University of Thessaloniki, 541 24 Thessaloniki, Greece
emanthos@civil.auth.gr

Abstract. The usage of certain amount of reclaimed asphalt (RA) is a well-established trend, which can be seen in many developed countries. Regrettably there are some countries or regions which still have very strict standard restrictions about RA usage in asphalt mixtures, mainly in surface layers. For example in current technical standards in the Czech Republic the use of RA in stone mastic asphalt mixtures or asphalt concrete for premium surface layers is banned. Recycling of construction materials saves direct primary costs like new material purchasing and at the same time also facilitates savings of other costs like energy, material disposal and transport etc. Those factors are connected to environment-friendly approaches, primarily in the case of emission and green-house gas production reduction. Material disposal can subsequently be reduced solely to waste or by-product materials which have no way of further utilization.

In this respect a 5 km trial section was designed including 12 different variants of stone mastic asphalt mixtures. The mixtures contained 0%, 20%, 30% and 50% reclaimed asphalt. The mixtures differed not only in amount of used RA, but also in source of RA (normal and SMA RA), used bituminous binder and used additives. The selected additives were based on cellulose fibers modified by an oil-based rejuvenator and crumb rubber modification in some alternatives of the used rejuvenator.

The performance of the asphalt mixtures was assessed by term of stiffness modulus (IT-CY method), thermal induced crack propagation (SCB test), flexural strength (three-point bending test) and resistance to permanent deformations. The key findings are presented in this paper.

1 Introduction

Stone Mastic Asphalt (SMA) mixtures are a specific aggregate skeleton based type of hot asphalt mixture used in surface layers mainly for heavy-loaded pavements. SMA mixtures were established in 1960s in Germany and for their properties they were very soon widespread to other countries. Nowadays SMA mixtures are one of the most used mixture in surface layer on motorways or intensively trafficked trunk roads. SMA

S. El-Badawy and R. Abd El-Hakim (Eds.): GeoMEast 2018, SUCI, pp. 158–169, 2019.
https://doi.org/10.1007/978-3-030-01908-2_13

mixtures are characterized by discontinuous granularity and higher bitumen content. The mixture has a coarse aggregate skeleton, which provides good stone-on-stone contact between the particles. The design of mixtures provides increased fatigue resistance, high resistance to permanent deformations and fatigue cracking as well as good skid resistance (*Sarang*).

Due to high bitumen content and discontinuous granularity it is necessary to prevent the bitumen drainage. The most frequently used additive for drainage prevention are cellulose or mineral fiber. In recent years there were research works pursuing to use of waste materials such as crumb rubber as an alternative material preventing bitumen drainage (*Mashaan*, *Sheng*). The crumb rubber can according to these findings not only prevent the bitumen drainage but also improve the properties of asphalt mixture.

Use of reclaimed asphalt (RA) in road constructions is nowadays a standard in most countries that can be characterised as advanced with respect to road infrastructure development. Recycling of construction materials saves direct primary costs like new material purchasing, material disposal, transportation and others. Those factors are also connected to environment-friendly approaches, like reduction of emission and greenhouse gas production, disposal and quarry expansion etc. The use of RA material in the production of new asphalt pavements leads to considerable economic and environmental benefits by transforming waste materials into a valuable resource. Use of recycled material in surface layers of highways and motorways is still debatable in many countries. The researches proved that use of reclaimed asphalt, even its elevated content, do not have to have negative influence on asphalt mixtures properties. It is always necessary to used homogenous material and mostly it is beneficial or even needed to use additives rejuvenating the aged bitumen in reclaimed asphalt or to seek for special bituminous binders. Despite the provided tests, the Czech technical standards do not allow RA usage in the superior surface layers.

2 Asphalt Mixture Variants

In the presented research study there is a summary of analysis of 12 control asphalt mix variants used in a trial section. The trial section was realized in autumn 2016 on a regional rural road with a daily average of 934 heavy load vehicles (*Czech national traffic counting*) and its total length was almost 5 km. On the trial section 18 different types of stone mastic asphalt SMA 11S with variable amount of RA was used. For this paper 12 of them are presented. The asphalt mixture variants SMA #1–#6 were not provided by the manufacturer, so the data are missing.

The SMA 11S mixtures differed in the percentage of used reclaimed asphalt (RA) of fraction 0/11 mm, the origin of applied RA (common RA from various asphalt concrete mixtures and SMA RA gained strictly from SMA layers) and the used additives (rejuvenator or additive based on rejuvenator and crumb rubber). For easier evaluation of influence of given additives and RA content, the mixture variants were divided into logical groups according to percentage of used RA. The “S” stays for superior asphalt mixture.

3 Input Materials

In the experimental study three types of additives were used in the presented asphalt mix variants. All of the additives should improve the properties of asphalt mixture containing elevated or high percentage of RA. The used additives were:

- cellulose fiber combined with special crumb rubber – labelled as “RE” – asphalt mix variant SMA #7;
- cellulose fiber modified by rejuvenator – labelled as “RF” – amount of rejuvenator in cellulose fiber granules varied according to the content of used RA – asphalt mix variants SMA #10–#13;
- crumb rubber modified combined with rejuvenator – labelled as “SE” – asphalt mixture variants SMA #14–#18.

Mixtures SMA #8 and #9 combine the cellulose fiber “R” and the crumb rubber modified combined with rejuvenator “SE”. The cellulose fiber “R” is commonly used in the SMA mixtures for prevention of bitumen drainage from aggregate. This fiber is not supposed to improve the properties of the mixtures with high RA content. The shortcuts in labelling of the mixtures were used for avoidance of commercial trade names or trademarks (Vacková et al. 2016).

Asphalt mixtures were designed with the most possible similar granularity and the production and paving of trial section was done in one day.

4 Testing of SMA Asphalt Mixtures

For evaluation of RA content and additive influence the following tests were performed and assessed on all mix variants:

- **volumetric characteristics** (CSN EN 12697-5, CSN EN 12697-6, CSN EN 12697-8);
- **stiffness modulus** (CSN EN 12697-26, method IT-CY) at test temperature of 0 °C, 15 °C a 27 °C;
- **resistance to permanent deformation** (CSN EN 12697-22) on small test device in air bath at test temperature of 50 °C;
- **flexural strength** determined by three point bending beam test at a test temperature of 0 °C (Czech Technical Conditions TP 151);
- **fracture toughness** and fracture energy determination according to SCB test (CSN EN 12697-44) performed on 100 mm diameter semi-cylindrical specimens.

4.1 Volumetric Characteristics of SMA 11S Mixtures

The volumetric characteristics were compared with requirements defined in product standard CSN EN 13108-5 (National Czech appendix). The air voids content limit required for control mixtures is 2–6%-vol. This limit was exceeded by two of the tested variants. Two asphalt mixtures with 20% RA (SMA #8 and #10) exceeded the upper standard limit. Although the air voids content of the third variant with 20% RA

(SMA #15) lays in the standard limits, the value is in comparison to all other variants noticeably higher. It means that three variants with 20% RA reached the three highest air voids contents. There are used different bituminous binders and different additives and so the results shows probable poor mix design. The granulometric composition was probably not optimally designed and it lead to air voids content increase.

The air voids content of asphalt mixtures and therefore the asphalt layer significantly influences the strength and deformation properties. If the air voids content is too high, the mixtures ordinarily have worsen strength properties, including stiffness modulus and water and frost resistance. On the other hand if the air voids content is to low, the mixtures have ordinarily problems with resistance to permanent deformation. Too dense mixtures is usually susceptible to rutting.

There is no visible influence of used additives in range of volumetric characteristics. They were more likely influenced by the granulometric composition and heterogeneity of used RA (Fig. 1).

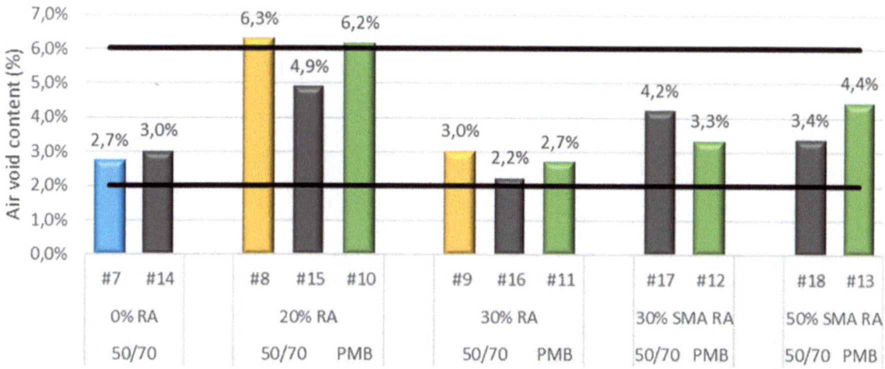

Fig. 1. Air void content of SMA 11S

4.2 Stiffness Determination

Stiffness modules were determined by non-destructive method of repeating loading in indirect tensile (IT-CY method according to EN 12697-26) at temperature of 0 °C, 15 °C and 27 °C.

The SMA 11S asphalt mixtures are solely used in wearing courses and the stiffness should not be considered as a key characteristic, however it can be used as a simple and fast indicator of the resistance of the mixture against deformation effects (Table 1).

The key temperature in Czech road design manual is 15 °C. Stiffness determined at 15 °C does not follow any expected trend. It would be expected that the stiffness modulus will increase with increasing RA content in asphalt mixtures. If higher RA content is used in the mixtures, it is assumed that the blended bituminous binder (binder from RA and virgin binder) will be stiffer and so the stiffness modulus of the asphalt mixture will be higher. But in these cases the rejuvenators were used, the rejuvenator softens the aged bituminous binder in RA and can cause the decrease of stiffness modulus (Fig. 2).

Table 1. Stiffness modulus of SMA 11S

		Air voids content	Stiffness modulus (MPa)			Thermal susceptibility
		(%)	0 °C	15 °C	27 °C	S_0/S_{27}
#7	0% RA + S-CEL RE	2,7%	17 827	9 148	3 264	5,5
#14	0% RA + Storelastic	3,0%	18 212	6 760	2 490	7,3
#8	20% RA + S-CEL 7G	6,3%	14 747	6 665	2 159	6,8
#15	20% RA + Storelastic	4,9%	16 861	6 093	1 917	8,8
#10	PmB + 20% RA + S-CEL RF 20	6,2%	12 115	3 914	1 486	8,2
#9	30% RA + S-CEL 7G	3,0%	16 656	7 806	2 768	6,0
#16	30% RA + Storelastic	2,2%	14 577	7 603	2 843	5,1
#11	PmB + 30% RA + S-CEL RF 30	2,7%	14 244	4 179	1 598	8,9
#17	30% SMA RA + Storelastic	4,2%	15 571	7 279	3 054	5,1
#12	PmB + 30% SMA RA + S-CEL RF 30	3,3%	13 967	5 065	1 766	7,9
#18	50% SMA RA + Storelastic	3,4%	15 004	7 077	3 029	5,0
#13	PmB + 50% SMA RA + S-CEL RF 50	4,4%	14 171	5 412	2 042	6,9

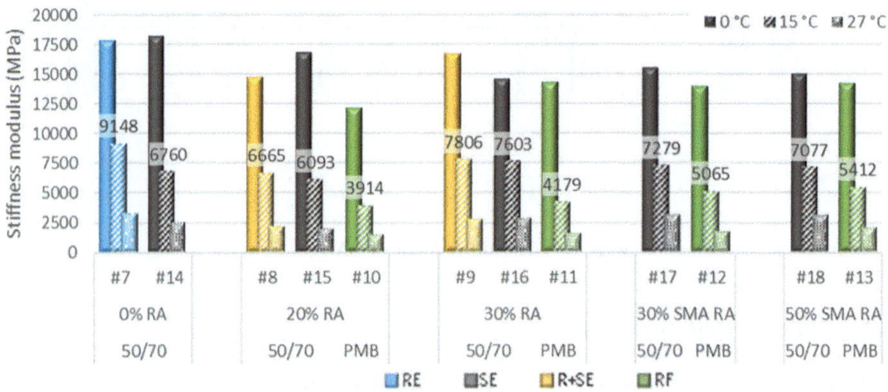

Fig. 2. Stiffness of SMA mix variants

The asphalt mix variant with cellulose fiber and special crumb rubber "RE" (SMA #7) reached significantly highest values. Using common paving grade binder 50/70 (even in combination with crumb rubber) the determined stiffness above 9000 MPa at 15 °C is beyond expectations. High stiffness modulus can predict good resistance to permanent deformation, but to certain extent can cause problems in range of low temperatures and related thermal induced cracking.

The mix variants with cellulose fibre combined with a rejuvenator RF (SMA #10–#13) reached significantly lower stiffness than the rest of the mixtures. The variants SMA #10 and #11 did not reach even half of the stiffness comparing to SMA #7 variant. The variant SMA #10 had an air voids content above the standard limit, but the variant SMA #11 had air voids content very close to the lower standard limit. It means, that the stiffness modulus decrease is not cause by the air voids content but by the effect of rejuvenator, which is softening the binder.

The mix variants SMA #14–#18 contained additive which combined crumb rubber and a rejuvenator "SE". Even here a decrease is apparent if comparing to SMA #7, but it is not as evident as in the previous case of used additive (decrease by 20–30%).

The mix variants SMA #8 and #9 contained not only the combined additive crumb rubber and rejuvenator "SE", but also common cellulose fiber "R". Adding of this fibre did not have any influence on stiffness at the test temperature of 15 °C, the mix variants with the same RA content reached basically the same values.

At the lowest test temperature (0 °C) the variant without RA with "SE" additive (SMA #14) reached higher stiffness than the variant with "RE" additive (SMA #7). The variants with "RF" additive again showed the lowest stiffness values. The trend seen at 15 °C, where conventional cellulose fibre "R" did not affect the stiffness, was not confirmed at the low test temperature. In this case, for mixture with 20% RA the stiffness decreased, but in the case of 30% RA there is on the other hand a significant increase.

The last used temperature was 27 °C. In this case the highest and lowest determined stiffness values were the same as at previous temperatures. Additional cellulose fiber "R" improved the property with 20% RA, but a significant effect in case of combining this fibre with 30% RA has not been proven.

4.3 Determination of Fracture Toughness (SCB Test)

Fracture toughness (characteristic used to describe the resistance to cracking) was determined based on the requirements of technical European standard EN 12697-44. The test specimens of 100 mm diameter were used instead on 150 mm described in the standard because of long-term assessment which is performed by the university for more than 5 years. To secure continuity in data sets the preference was kept to the smaller diameter also in this experimental study. Additionally, the standard requires the test specimens being compacted by gyratory compactor which was replaced by traditional impact compactor. For sake of this project the specimens used previously for determination of stiffness were used. The test specimens were cut on laboratory saw to gain the proper dimensions. The test procedure was performed at a test temperature of 0 °C applying a standardized loading rate of 5 mm/min (Fig. 3).

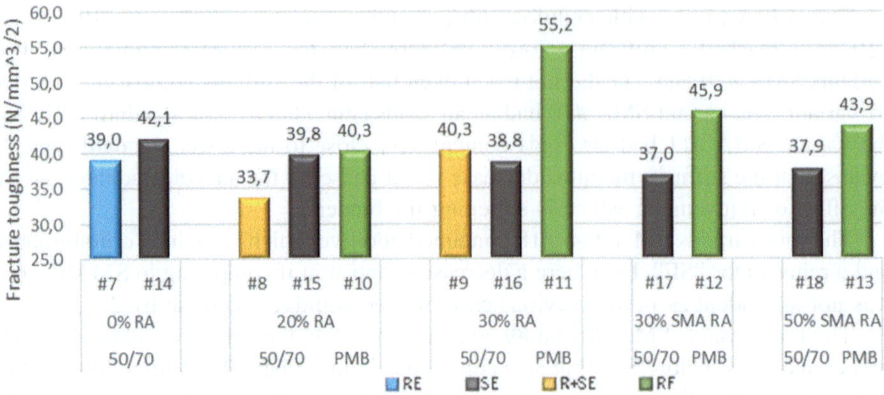

Fig. 3. Fracture toughness of SMA mix variants

The essence of the test is a three-point bending using semi-cylindrical specimens with a defined notch in the middle of bottom specimen face. If the stiffness of the specimens is too high, the fracture toughness is usually decreasing showing the relationship between stiffness and brittleness of a viscoelastic material. The stiff specimens are expected to be more predestined to brittle cracking.

Results of fracture toughness to same extent correspond with the results of stiffness modulus – the higher stiffness, the lower fracture toughness. The mixtures with cellulose fiber "RF" showed the highest fracture toughness values, with exception of the variant SMA #10, where a higher drop is visible. This phenomenon can be cause by the fact, that mixtures with 20% of RA (SMA #8, #10 and #15) had very high air voids content, which negatively influences the behaviour in the low temperature range.

The mix variant without RA (SMA #7 and #14) reached very high stiffness, but simultaneously still demonstrated quite good result of fracture toughness. These results indicates important influence of virgin (unaged) bituminous binder on properties of asphalt mixtures.

4.4 Resistance to Permanent Deformation

Resistance to permanent deformation was determined according to the standard EN 12697-22 in small testing device and in air bath at temperature of 50 °C. The maximum allowed value of proportional rut depth (PRD_{AIR}) is according to the product standard EN 13108-5 given by 5% and the maximum value of wheel tracking slope (WTS_{AIR}) cannot exceed 0.07 mm.

All tested variants comply with the limits given for SMA mixture in the specifications. The result of resistance to permanent deformation are more likely comparable to each other with exception of mixture SMA #16. All mixtures reached values of PRD_{AIR} in range of 2.3–3.2%. The variant SMA #16 exceeded 4.0%. This variant had an air voids content on boundary of the lower limit required by the specifications and as it was stated earlier – the mixtures with very low air voids contents have usually problems with fulfilling acceptable value of rutting (Figs. 4 and 5).

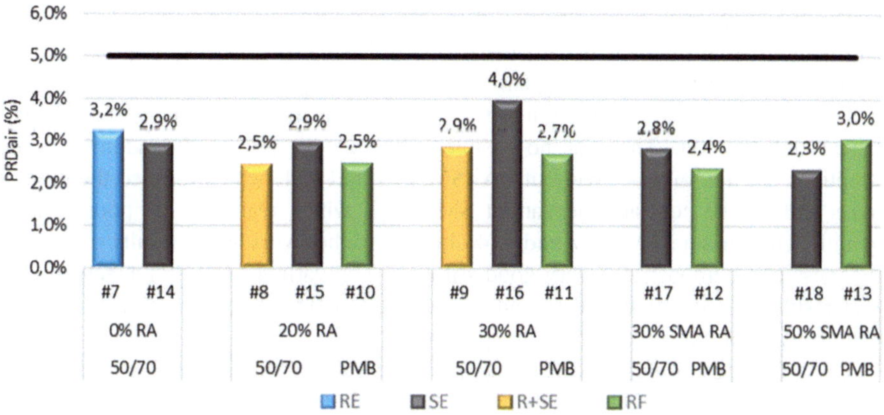

Fig. 4. Resistance to permanent deformation – PRD_{AIR} parameter

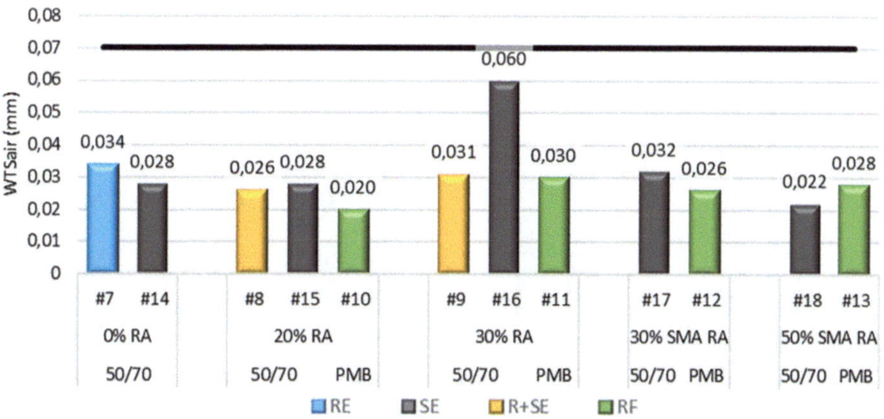

Fig. 5. Resistance to permanent deformation – WTS_{AIR} parameter

There is no visible coherent influence of used additives on rutting properties of the asphalt mixture. On the other hand it is important to emphasize that mix variant which combines paving grade binder, RA and additive “SE” (combination of crumb rubber and rejuvenator) always resulted in the highest PRD_{AIR} values in the test group where same RA content was used. All mixtures fulfilled the standard criterion, but it is always necessary to evaluate the particular influence of the given additive.

4.5 Flexural Strength – Three Point Bending Test

The tensile bending strength was determined in compliance with the Czech Technical Specifications (Ministry of Transportation Czech Republic, 2010) TP 151. This methodology is primary used for asphalt mixtures with high stiffness modulus (HMAC), but due to its simplicity and easy reproducibility it was applied even on SMA mixtures. The essence of the test is three point bending of an unnotched beam (50 × 40 × 320 mm) at a temperature of 0 °C with loading rate of 1.25 mm/min.

The test parameters were extended above the range of required technical conditions and fracture energy was calculated as well. This characteristic is calculated as the area under the force-displacement diagram.

The results of flexural strength are balanced as in the case of resistance to permanent deformation. There is no visible influence of any of the additives. The exceptions are the mixtures without RA (SMA #7 and #14) which reached the highest values and on the contrary the variant SMA #12, which reached very poor flexural strength. The mixture SMA #7 and #14 already reached very good results of fracture toughness and demonstrated very good stiffness. The variant SMA #12 reached quite good results in case of fracture toughness characteristic, so the decrease in flexural strength was not expected and is unexplainable.

These findings could indicate some less favourable influence of RA in the compared asphalt mixtures. On the other hand the results of most evaluated asphalt mixtures were always in standard boundaries and were in many cases very good. The SMA #12 variant reached quite good results of fracture toughness, so the decrease in flexural strength is expected and unexplainable (Figs. 6 and 7).

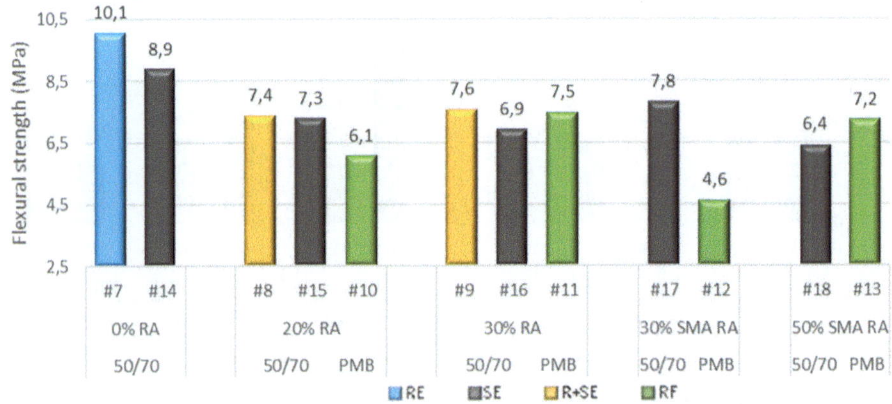

Fig. 6. Three point bending – Flexural strength

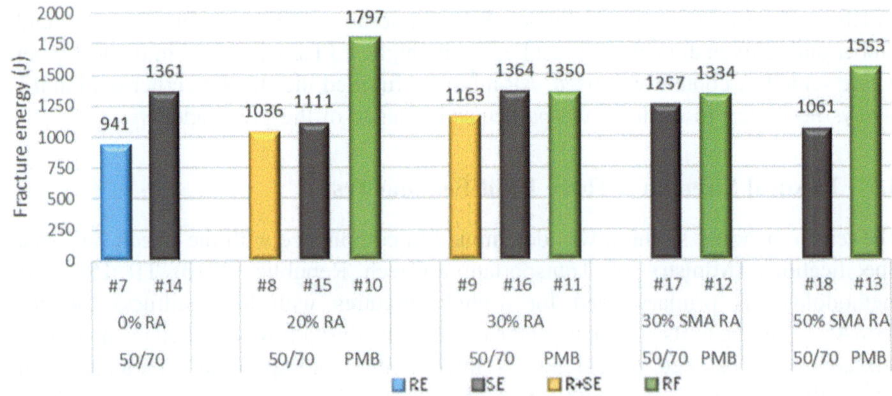

Fig. 7. Three point bending – Fracture energy

5 Conclusions

Within this research study, some basic as well as more performance related properties of 12 asphalt mix variants sampled from a trial section were determined. The variants were divided into 5 logic groups according to their RA content and were further evaluated. Tested additives were cellulose fiber containing special crumb rubber “RE”, cellulose fiber modified by a rejuvenator “RF” and crumb rubber combined with a rejuvenator “SE”.

Rejuvenator in asphalt mixtures containing reclaimed asphalt influences to certain level the workability of the mixtures and has positive effect on air voids content. The highest air voids content were found for mixtures containing 20% RA. Unfortunately in this case the problem was probably a poor granulometric design. Two of the variants (SMA #8 and #10) exceeded the standard limit, the third was below the limit provided by the specifications, but still very high. The additive did not seem to noticeably influence the air voids content. The volumetric properties were mostly influenced by the granulometric design.

Rejuvenator should soften the bituminous binder in the asphalt mixture and therefore cause to certain level the decrease of strength properties (if they would be compared to “reference” mixture with the same RA content and without any additive).

Asphalt mix variants with cellulose fibre and a special crumb rubber “RE” reached the highest values of stiffness as same as highest values of flexural strength. Fracture toughness of this variant was in the middle interval of gained results, but the depth of the rut was slightly higher than at shown for the remaining mix variants. The resistance to permanent deformation was deeply under the limit required by the specifications, but in comparison to other variants the mixture SMA #7 evinced slightly worse result.

Another used additive was cellulose fibre modified by rejuvenator “RF”. Rejuvenator was dosed into the fibre according to the final intended use – the higher RA content was designed in the asphalt mixture, the higher amount of rejuvenators was in the cellulose fibre granules. The used rejuvenator generally decreases considerably the strength properties. This was visible even in this experimental study. The mixture variants reached the lowest stiffness and lower (but not lowest) flexural strengths. For stiffness, there is a clear trend showing that if RA content is increasing the stiffness increases as well. The difference between mixtures with 20% and 50% reclaimed asphalt is almost 30%. For resistance to permanent deformation it is no possible to unambiguously determine the influence of this additive, but the mixtures generally reached lower ruts. The most positive influence is apparent at the fracture toughness, when mixture with additive “RF” reached the highest values for this parameter. This partially corresponds with the fact of low stiffness modules – the more stiff mixture, the worse behaviour in the range of low temperatures.

Last additive used in this study was crumb rubber combined with a rejuvenator “SE”. For variants (SMA #8 and #9) additionally common cellulose fibre “R” was used. Due to the fact that not only higher content of RA was used, but also the crumb rubber is used in the asphalt mixture, the mixtures do not behave homogeneously in all cases. The stiffness values of variants with “SE” are higher than in case of “RF” additive, but lower than gained for “RE” additive. Only the mixture without RA reached for some test temperatures values similar to results of variants containing additive “RE”. Fracture toughness of these variants is lower, than it was expected and the lowest value (SMA #11) achieved only 60% of the highest fracture toughness. Additional usage of cellulose fibers “R” did not demonstrate any significant positive effect on the mix properties. Only resistance to permanent deformation was in this case considerably better.

The presented study proved that even SMA 11S mixture with 50% RA can fulfil the limits required by the product standard EN 13108-5 (national appendix). Even though the SMA mixtures are used in wearing courses, where they are exposed to climate impacts and changes as well as to direct contact with the traffic loading, are not required to be tested for thermal induced cracking or in terms of water susceptibility (ITSR). Adding of at least these performance requirements in the specifications would help the manufacturers to push forward the usage of higher content of RA in asphalt mixtures.

Acknowledgements. This paper was elaborated within the activities of the project no. TA04031328 as supported by the Technology Agency of Czech Republic.

References

Czech National Traffic Counting: http://scitani2016.rsd.cz/pages/intenzitytable/default.aspx?s=1-4410

Mashaan, N.S., Ali, A.H., Koting, S., Karim, M.R.: Performance evaluation of crumb rubber modified stone mastic asphalt pavement in malaysia. Adv. Mater. Sci. Eng. 1–8 (2013). https://doi.org/10.1155/2013/304676. Accessed 25 Apr 2018. ISSN 1687-8434

Mashaan, N.S., Karim, M.R., Aziz, M.A., Ibrahim, M.R., Katman, H.Y., Koting, S.: Evaluation of fatigue life of CRM-reinforced SMA and its relationship to dynamic stiffness. Sci. World J., pp. 1–11 (2014) (Cite 2018–04–25). https://doi.org/10.1155/2014/968075. ISSN 2356-6140

Ministry of Transportation Czech Republic: Technical specifications: TP 151 – High modulus asphalt mixtures, Prague, (only in Czech) (2010)

National Standard CSN EN 13108-5: Bituminous Mixtures - Material Specifications - Part 5: Stone Mastic Asphalt

National Standard National Standard CSN EN 12697-6: Bituminous Mixtures - Test Methods - Part 6: Determination of Bulk Density of Bituminous Specimens

National Standard CSN EN 12697-5: Bituminous Mixtures - Test Methods - Part 5: Determination of the Maximum Density

National Standard CSN EN 12697-8: Bituminous Mixtures - Test Methods - Part 8: Determination of Void Characteristics of Bituminous Specimens

National Standard CSN EN 12697-26: Bituminous Mixtures - Test Methods - Part 26: Wheel Tracking

National Standard CSN EN 12697-26: Bituminous Mixtures - Test Methods - Part 26: Stiffness

National Standard CSN EN 12697-44: Bituminous Mixtures - Test Methods - Part 44: Crack Propagation by Semi-circular Bending Test

Vacková, P., Valentin, J., Blažek J.: Selected Characteristics for Variants of SMA Asphalt Mixtures from the Trial Section Kněževes – Rakovník. Research report for the year 2016, Czech Technical University in Prague, (only in Czech) (2016)

Valentin, J., et al.: Asphalt Mixtures of SMA Type Using Different Types of Cellulose Fibers and Reclaimed Asphalt Material. Research report for the year 2016, Czech Technical University in Prague, (only in Czech) (2015)

Sarang, G., Lekha B.M., Geethu J.S., Ravi Shankar A.U.: Laboratory performance of stone matrix asphalt mixtures with two aggregate gradations. J. Mod. Transp. 130–136 (2015). https://doi.org/10.1007/s40534-015-0071-5. Accessed 21 Mar 2018. ISSN 2095-087X

Sheng, Y., Li H., Guo P., Zhao G., Chen H, Xiong R.: Effect of fibers on mixture design of stone matrix asphalt. Appl. Sci. (2017). https://doi.org/10.3390/app7030297. Accessed 19 Mar 2018. ISSN 2076-3417

The Applicability of Recent Mechanically Stabilized Granular Layer Concept in ME Pavement Design

Zikmund Rakowski[1(✉)], Jacek Kawalec[2,3], and Sławomir Kwiecień[3]

[1] Consultant, Český Těšín, Czech Republic
zyraki44@gmail.com
[2] Tensar International Sro, 73701 Český Těšín, Czech Republic
[3] Silesian University of Technology, Gliwice, Poland

Abstract. Recent concept of the structure and the mechanics of mechanically stabilized granular layers is described from the point of view of pavement design. Described approach is applied in finite elements model of an example of typical pavement structure. Influence of mechanical properties of mechanically stabilized layer on the range of the strain on the base of asphalt layer is analysed. The results are implemented into pavement performance models using various empirical equations. The consequence on serviceability of the pavement structure is shown. Mechanically stabilized granular layer concept is proven as an effective approach to pavement design.

1 Introduction

Layers of granular materials are typical construction elements of paved and/or unpaved roads, railways and hard standings structures. The granular layers experience only the relatively insignificant permanent load from the weight of the layers on top. When construction or in-service traffic passes (vehicles, trains, etc.), these layers are under transient increased pressure resulting in associated strain. Passing of the trafficking load, induces both vertical and horizontal stress components in granular layers (Kwon et al. 2008; Byun and Tutumluer 2018; Robinson et al. 2018), where the associated horizontal strains have a radial distribution. The horizontal components of stress cause the particles in the layer to move laterally and after the traffic has passed they do not recover completely to the condition before the load was applied. Cumulative lateral movement after numerous of loading cycles cause permanent deformation and will eventually result in a condition beyond serviceability of the layers (Cook et al. 2016; Jenner et al. 2002; Tutumluer et al. 2012; Grygierek and Kawalec 2017). In this respect the stabilization is defined as the beneficial consequence on the serviceability of an unbound granular layer via the inhibition of the movement of the particles of that layer under applied load (EOTA 2012; Belyaev 2007; Zornberg 2017; Hornicek and Rakowski 2018).

S. El-Badawy and R. Abd El-Hakim (Eds.): GeoMEast 2018, SUCI, pp. 170–180, 2019.
https://doi.org/10.1007/978-3-030-01908-2_14

2 Actual Concept of Mechanically Stabilized Layer (MSL)

Key mechanism involved in the stabilization is interlock of aggregate provided by a stiff geogrid structure (Kwon and Tutumluer 2009; Qian et al. 2018). Effect of geogrid could be then model in DEM approach (Konietzky 2002). During the placement and compaction of a granular layer over a geogrid, the aggregate particles partially penetrate into the apertures and abut against the ribs of the geogrid. This results in interaction of the geogrid and the aggregate under applied load by locking of the aggregate grain in the aperture. In this way the effect of interlock is defined as the inhibition of the movement of the particles of granular layer under applied load. In this respect the first level of the observation of that mechanism is the scale of the stone grain (Fig. 1).

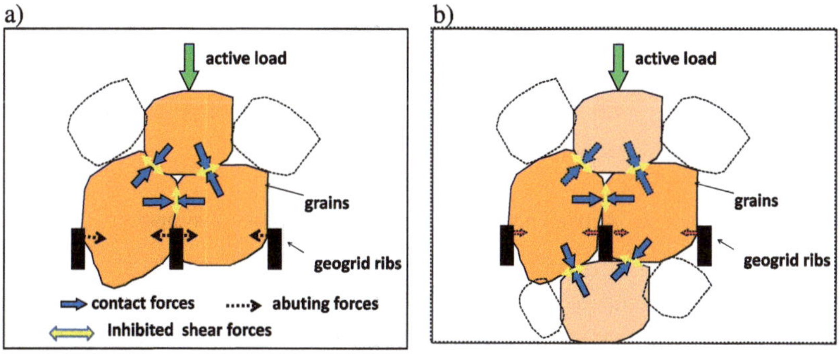

Fig. 1. Stone grains immobilized in geogrid aperture and grain above it (a) and above and below it (b)

The scheme on Fig. 2 show that the stone grain is properly locked in the aperture and when loaded abutting forces are becoming active. The grain in the next row acts down by contact forces between grains and by that is locked as well. In case that geogrid is located within granular layer also grain in the row below is locked by contact forces (Fig. 1b). Shear displacements between grains are restricted almost to zero. The system is stiff and performs closely as "quasi rock" (immobilized grains form a kind of small piece of massive rock). Important features of geogrids are described by Giroud (2009): grain/aperture size ratio, geogrid rib shape and stiffness, aperture shape and stiffness, grain/aperture size ratio is commonly related to D_{50} index. But field observations show that when the granular material is properly installed (by cascading not pushing) the larger stones are first on geogrids simply due to gravitation. In this respect then D_{max} should be the decisive parameter here. The range of grain size varies from ca. 110% of the aperture size (100% and less would mean moving through the aperture without abutting) up to $1.4 \times d_g$ (aperture size). It is derived from the geometry of abutting forces, when considering the grain is larger than $1.4 \times d_g$, vertical lifting component is becoming greater than horizontal abutting component so the locking is much weaker. Further according to Giroud (2009); Giroud and Han (2016) the stiffer geogrid rib the better, the thicker rib the better, vertically shaped rib is better than horizontal one. An attention should be paid also to the second row of grains over those penetrated into grid aperture. Such a grain is exposed to abutting forces from the bottom grains (when active load is applied).

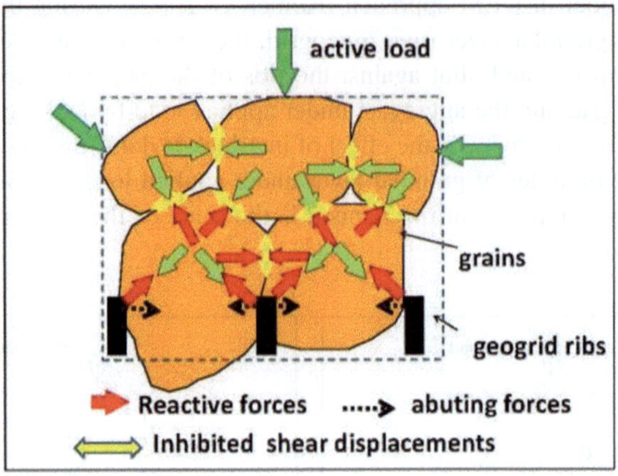

Fig. 2. Active and reactive forces in interlocking system

The reality is that up to now does not existed any method for measuring the immobilization effect in geogrid aperture directly. Methods being used so far are often sophisticated but still rather indirect (Tutumluer 2016; Zornberg 2016) (shear wave propagation, pull- out test, respectively). Interesting approach was published in Matys and Baslik (2004) when cone similar to CPT test one was pushed through granular material, with different geogrids and without any. Pushing forces at various distances from the geogrid were recorded, recognizing well the resistance of immobilized grains against pushed cone. Ideally the mechanism should be described by recording the acting force and resulting displacement of the grains. Important feature is that it is not any more discrete granular material and geogrid as separate elements but a kind of composite with new quality performing like stiff continuum. It means that macro deformations typical for granular material under load are replaced here by micro deformations similar like in case of rock. Basic unit of that composite can be described as a stiff triangular body (pyramid). In plane continuous layer of stiff pyramids is formed (Fig. 4a). Such a layer has specific mechanical properties: it is relatively thin comparing with the area of the layer and is stiff and strong in plane direction. The efficiency of interlocking is critical here because if the displacement of the grain is still somehow possible, then proper stiffening can not be achieved like e.g. in case of flexible geogrids (Matys and Baslik 2004). The performance of properly immobilized continuum layer means that resistance against particle movement (immobilization) is mobilized in rather wide area of the layer (Fig. 4b). In case of hexagonal geogrids that stiffening is rather homogenous in all in-plane directions resulting in regular radial distribution of load (Fig. 4c).

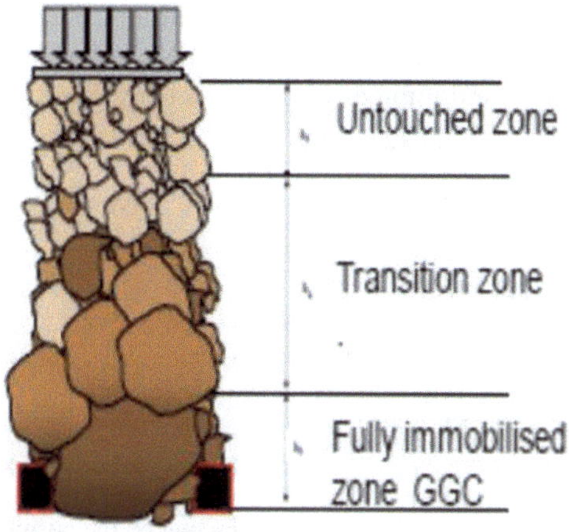

Fig. 3. Zonality of the MSL

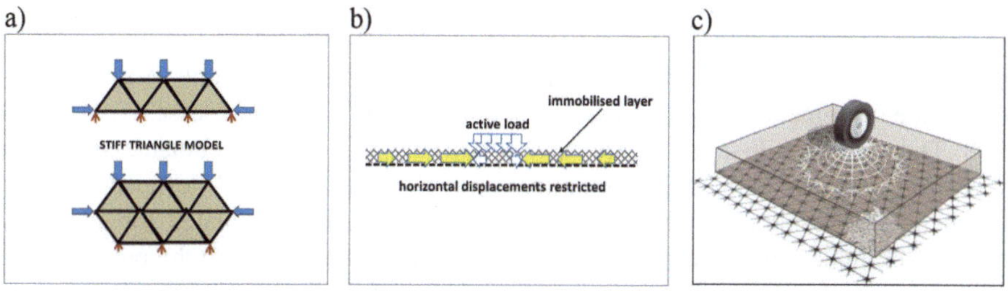

Fig. 4. Triangle model of immobilized layer (**a**), stiffening effect in its plane (**b**), radial distribution of load (**c**)

Immobilized/stiffened layer should be characterized namely by its thickness, deformability (modulus) both in in-plane direction and perpendicularly as well, and shear strength. Several researchers, e.g. Konietzky et al. (2004); Stahl et al. (2014); Fischer and Szatmari (2016), confirmed that fully immobilized zone can achieve the thickness h_i = 7–10 cm. In case of hexagonal geogrid it can be even up to 15 cm, it means several times D_{max}, due to firmer abutting of grains above the geogrid level. Then, logically the effect of abutting forces is decreasing with the distance from the geogrid (Fig. 3).

It means that the layer mechanically stabilized by hexagonal geogrid is vertically not homogenous from the point of view mechanical immobilization of grains. Three different zones can be distinguished: immobilized/stiffened zone $\mathbf{h_i}$, transition zone $\mathbf{h_T}$, not influenced zone $\mathbf{h_o}$. Obviously the performance of the whole MSL depends on the intensity of immobilization of the base zone, then thickness of zones and proportion of $\mathbf{h_i}$:$\mathbf{h_T}$: through the whole structure of the MSL. The properties of $\mathbf{h_i}$:$\mathbf{h_T}$ zones will be different for different geogrids and adjacent granular material. For the certain type of the geogrid and its interaction with specific granular material the properties can be treated as typical for such defined zones. Thickness of the zone $\mathbf{h_o}$ and its share in total MSL will play very important role. The structure of MSL depends also on the position of the geogrid in the layer and number of geogrids (Fig. 5).

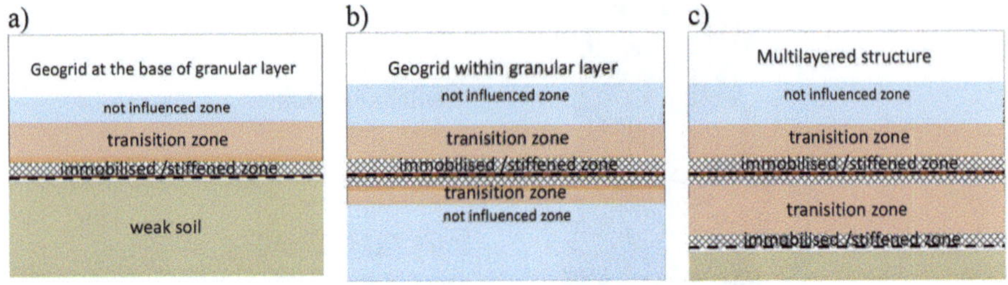

Fig. 5. Zonality of mechanically stabilised layer depending on the position (**a**, **b**) and number of geogrids (**c**)

3 Mechanical Empirical Pavement Design (Mepd)

Pavement design using the MEPD is an iterative process—the outputs from the procedure are pavement distresses and smoothness, not layer thicknesses (AASHTO 2008).The designer first considers site conditions (i.e. traffic, climate, subgrade, existing pavement condition for rehabilitation) in proposing a trial design for a new pavement or rehabilitation strategy. The trial design is then evaluated for adequacy against user input, performance criteria, and reliability values through the prediction of distresses and smoothness. Duplication is a violation of applicable law. the design does not meet the desired performance criteria at the specified reliability, it is revised and the evaluation process repeated as necessary. According to (AASHTO 2008) resilient modulus is required for all unbound paving layers and the foundation but geogrids cannot be simulated in the MEPD at this time.

The situation is different now having on mind knowledge on mechanically stabilized layers (MSL) as described in previous chapter. On the flow chart (Fig. 6) we can see the place of MSL properties in the working scheme for the design. Executed experimental works (Rakowski 2017) enabled to derive basic mechanical properties as resilient modulus and Poisson ratio for the MSL. For fully immobilized zone of the MSL the values E = 1.5–2 GPa, ν = 0.01–0.02 were derived.

4 FEM Modelling

The aim of FEM modeling used for analysis shown in this paper was to demonstrate an influence of MSL mechanical properties on basic parameters of pavement layered structure. The structure of the model is shown on Figs. 7 and 8. Parameters of the layer 3 are variable to study their influence mainly on tensile strains on the base of asphalt layers. The strain there has fundamental influence on potential development of tensile cracking.

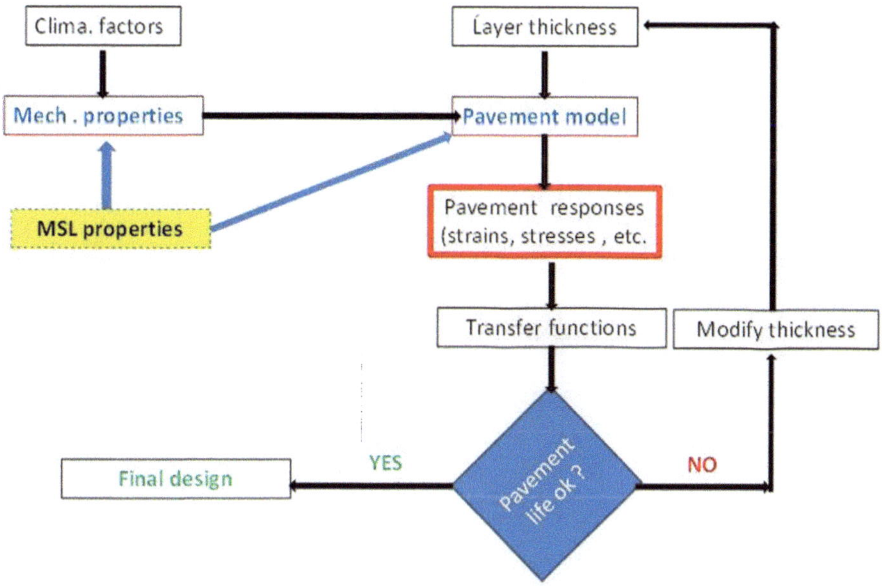

Fig. 6. Mechanistic empirical design flow chart (MEPD)

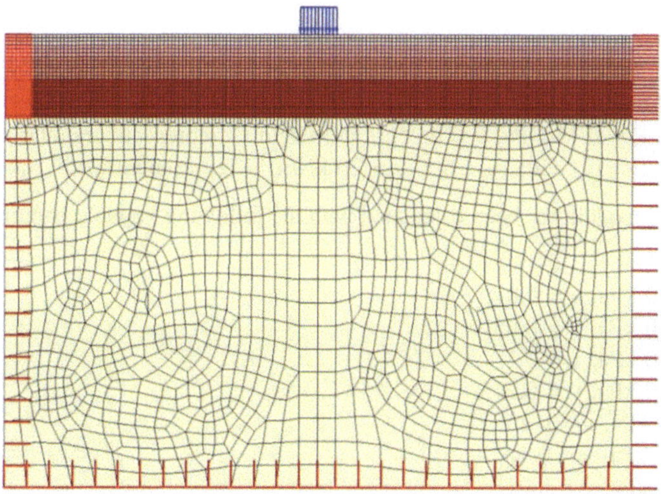

Fig. 7. General view of the FEM model (dimensions: 4.25 × 3 m)

Following parameters where implemented into model:

- Layer 1: E = 5 GPa, ν = 0.35, γ = 20 kN/m^3 - represents asphalt wearing course,
- Layer 2: E = 2 GPa, ν = 0.35, γ = 20 kN/m^3 - represents asphalt bearing course,

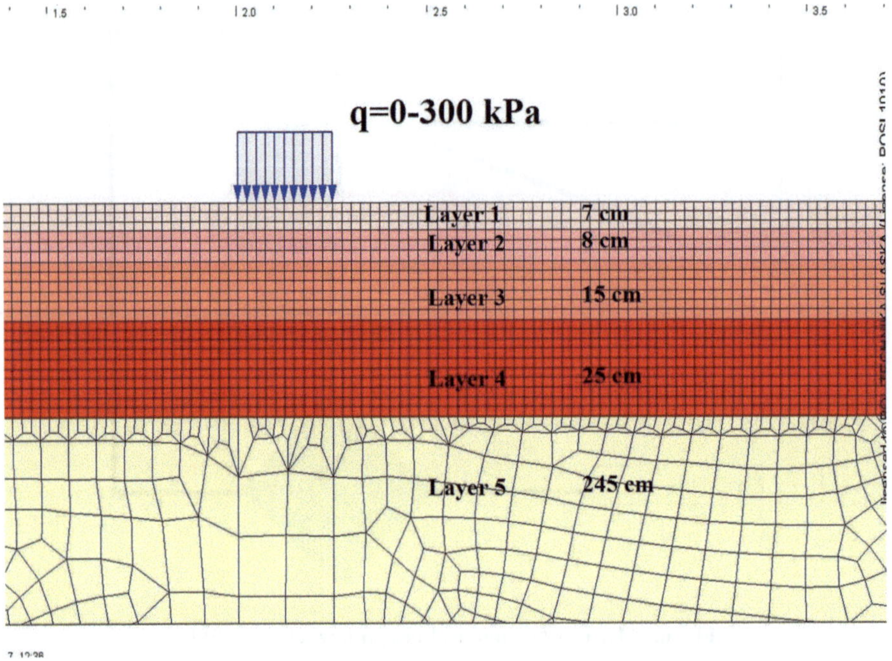

Fig. 8. Detailed view of the FEM model

- Layer 3: E = 0.25–1.5 GPa, ν = 0.01–0.25, γ = 20 kN/m^3- variable: represents standard granular base or MSL,
- Layer 4: E = 0.15 GPa, ν = 0.25, γ = 20 kN/m^3- represents granular subbase,
- Layer 5: E = 0.1 GPa, ν = 0.3, γ = 20 kN/m^3 - represents subsoil.

In total 7 models has been calculated with the following variable parameters for the layer 3:

1. E = 1500 kPa, ν = 0.01,
2. E = 1500 kPa, ν = 0.05,
3. E = 1500 kPa, ν = 0.1,
4. E = 1500 kPa, ν = 0.25,
5. E = 250 kPa, ν = 0.25,
6. E = 250 kPa, ν = 0.1,
7. E = 250 kPa, ν = 0.01.

5 Interpretation and Discussion of the Results

Horizontal strain on the top of granular and/or MSL layer underneath bearing asphalt course is the parameter with essential impact on fatigue cracking development from the base up to the surface. Therefore interpretation of the models was oriented on that namely. The horizontal strain maps of characteristic models 4 and 5 are on Figs. 9 and 10. The main conclusion from the analyze of those maps are:

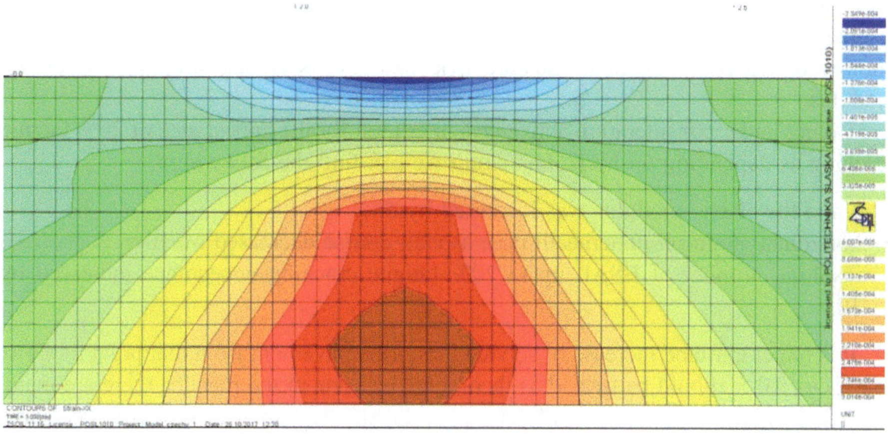

Fig. 9. Horizontal strains in the model 4

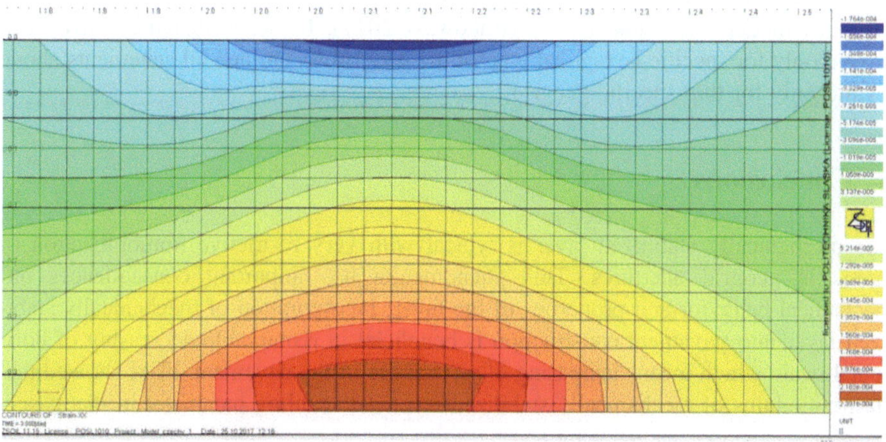

Fig. 10. Horizontal strains in the model 5

- Poisson ratio changes have no essential influence on horizontal strain on the base of asphalt layer (boundary with MSL),
- the main influence appeared when changing E modulus from standard 250 MPa for regular granular base to 1500 MPa representing MSL,
- the differences in strains on the base of asphalt layer (boundary with MSL), are of the order 10, e.g. Model 4 $\varepsilon_x = 6.7 \times 10^{-5}$, Model 5 $\varepsilon_x = 2.3 \times 10^{-4}$.

Appearance of fatigue cracking is fully empirical function. Expected strain is the result of FEM. There are various equations for determination of appearance of fatigue cracking as the function of the expected horizontal strain on the base of asphalt layer for 20% fatigue cracking appearance criterion, e.g. ($N_f = 0.0796(e_t)^{-3.291}\ (E^*)^{-0.854}$), - Asphalt Institute (1981) or ($N_f = 5 \times 10^{-6}(e_t)^{-30}$) – Illinois DOT (1992), where N_f is number of ESAL (equivalent standard axle load), e_t is tensile strain, E* is elastic modulus.

Taking under consideration above design criteria for the appearance of 20% fatigue cracking as depending on horizontal strain we can find the following:

- while the elastic modulus was increased by 6x,
- horizontal strain on underneath base asphalt course decreased by 10x,
- resulting number of ESAL up to 20% fatigue cracking appearance increased by 10^2x.

It is rather extreme increase of service life of the pavement from that point of view. The factor 10^2x should be treated as theoretical because in the practice there are many other factors influencing service life of asphalt layers. But anyhow, if it would be some multiplication of number of ESAL it is very worth to be considered. It can be stated that incorporation of the MSL into pavement structure has potentially very essential influence on service life of flexible pavements. But it must be emphasized that the proper function of the MSL is not generic for any type of the geogrid. The input data for FEM models were taken from (Rakowski 2017). It means that they have been proved only for specific hexagonal type of the geogrid so far, and all statements above a valid only for that specific geogrid.

6 Conclusions

The mechanically stabilized layer is typical with very restrictive immobilization of grains in geogrid apertures. This phenomenon has important consequences on mechanical behavior and properties of the layer. Incorporation of mechanically stabilized layer into pavement construction is beneficial in remarkable decreasing of horizontal strain which means increasing of service life of the pavement. FEM showed that elastic modulus of mechanically stabilized layer is the ruling parameter in that case.

Acknowledgements. The authors thanks Tensar International, Ltd., UK, for the support and financing experimental works.

References

AASHTO: Mechanical Empirical Pavement Design Guide, A Manual of Practice (2008)

Asphalt Institute: Thickness Design-Asphalt Pavements for Highways and Streets. Manual Series No. 1 (1981)

Belyaev, V.S.: Full scale experiment of the role of geosynthetics in railway structure over karst. Tensar International internal report, St.Petersburg, Russia (2007)

Byun, Y. H., Tutumluer, E.: Local stiffness quantification of geogrid stabilized aggregates in relation to deformation behavior. In GeoShanghai International Conference, pp. 109-114. Springer, Singapore (2018)

Cook, J., Dobie, M., Blackman, D.: The development of APT methodology in the application and derivation of geosynthetic benefits in roadway design. In: The Roles of Accelerated Pavement Testing in Pavement Sustainability, pp. 257-275. Springer, Cham (2016)

EOTA: Non-reinforcing hexagonal geogrid for the stabilization of unbound granular layers by way of interlock with the aggregate. EOTA Report TR41 (2012)

Fischer, S., Szatmari, T.: Investigation of the geogrid-granular soil combination layer with laboratory multi-level shear box test. In: Proceedings of EuroGeo 6, Conference Special Sessions on Geosynthetics in Road Construction, Ljubljana, Slovenia (2016)

Giroud, J.P.: An assessment of the use of geogrids in unpaved roads and unpaved areas. In: Proceedings of the Jubilee Symposium Polymer Geogrid Reinforcement, Institution of Civil Engineers, London, UK, pp. 23–36 (2009)

Giroud, J., Han, J.: Mechanisms governing the performance of unpaved roads incorporating geosynthetics. Geosynthetics **34**(1), 22–36 (2016)

Grygierek, M., Kawalec, J.: Selected laboratory research on geogrid impact on stabilization of unbound aggregate layer. Proc. Eng. **189**, 484–491 (2017)

Hornicek, L., Rakowski, Z.: The mechanically stabilized granular layers - an effective solution for tunnel projects. In: International Congress GEOMEAST 2018, Cairo, Egypt (2018)

Illinois Department of Transportation: Materials and Procedures for Sealing and Filling Cracks in Asphalt-Surfaced. Pavements, 0046-16, 11/15 (1992)

Jenner, C.G., Watts, G.R.A., Blackman, D.I.: Trafficking of reinforced, unpaved subbases over a controlled subgrade. In: Proceedings of the Seventh International Conference on Geosynthetics: State of the Art-Recent Developments, 7-ICG, Nice, France, vol. 3 (2002)

Konietzky, H.: Numerical modeling in micromechanics via particle methods. In: International PFC Symposium, Gelsenkirchen, Germany, 6–8 November 2002. CRC Press (2002)

Konietzky, H., te Kamp, L., Groeger, T., Jenner, C.: Use of DEM to model the interlocking effect of geogrids under static and cyclic loading. In: Numerical Modelling in Micromechanics via Particle Methods, Balkema, Leiden, Netherlands, pp. 3-12 (2004)

Kwon, J., Tutumluer, E., Al-Qadi, I., Dessouky, S.: Effectiveness of geogrid base-reinforcement in low-volume flexible pavements. In: GeoCongress 2008: Geosustainability and Geohazard Mitigation, pp. 1057–1064 (2008)

Kwon, J., Tutumluer, E.: Geogrid base reinforcement with aggregate interlock and modeling of associated stiffness enhancement in mechanistic pavement analysis. Transp. Res. Rec. J. Transp. Res. Board **2116**, 85–95 (2009)

Matys, M., Baslik, R.: Study of interlocking effect by the push test. In: Proceedings of the 3rd Asian Regional Conference on Geosynthetics GeoAsia2004, Seoul, pp. 341–348 (2004)

Qian, Y., Tutumluer, E., Mishra, D., Kazmee, H.: Triaxial testing and discrete-element modelling of geogrid-stabilised rail ballast. In: Proceedings of the Institution of Civil Engineers-Ground Improvement, pp. 1–22 (2018)

Rakowski, Z.: An attempt of the synthesis of recent knowledge about mechanisms involved in stabilization function of geogrids in infrastructure constructions. Proc. Eng. **189**, 166–173 (2017)

Robinson, W.J., Tingle, J.S., Norwood, G.J., Wayne, M.H., Kwon, J.: Performance of multi-axial geogrid stabilised flexible pavements. In: Proceedings of the Institution of Civil Engineers - Ground Improvement, pp. 1–32 (2018)

Stahl, M., Konietzky, H., Te Kamp, L., Jas, H.: Discrete element simulation of geogrid-stabilised soil. Acta Geotech. **9**(6), 1073–1084 (2014)

Tutumluer, E., Huang, H., Bian, X.: Geogrid-aggregate interlock mechanism investigated through aggregate imaging-based discrete element modeling approach. Int. J. Geomech. **12**(4), 391–398 (2012)

Tutumluer, E.: Geogrid-aggregate interlock mechanism investigated. Special session on "Geosynthetics in Road Construction" EuroGeo 6, Ljubljana, Slovenia (2016)

Zornberg, J.: Stabilization of roadways using geosynthetics. Special session on "Geosynthetics in Road Construction" EuroGeo 6, Ljubljana, Slovenija (2016)

Zornberg, J.G.: Functions and applications of geosynthetics in roadways. Proc. Eng. **189**, 298–306 (2017)

Alternative Modified Binders for Mastic Asphalt Layers

Jakub Šedina, Jan Valentin(✉), and Petr Mondschein

Department of Road Structures, Czech Technical University in Prague, Prague, Czech Republic
{sedinjak, jan.valentin}@fsv.cvut.cz, mondscheinp@gmail.com

Abstract. The ever-increasing requirements on quality and life-span of asphalt layers present a reasonable motivation to explore new approaches and technologies. For many years, the technical options have included modification of regular paving grade bituminous binders. This paper focuses on the possible use of bituminous binders used; both activated and chemically improved by selected modifiers, in mastic asphalt as well as on their potential beneficial effects on selected parameters of the mixtures tested. The modification of reference paving grade bitumen 20/30 involved micromilled (pulverized) recycled crumb rubber as well as a new generation of synthetic waxes or even combination of the two types of additives. Potential benefit can be seen in the improved deformation characteristics and in the possibility to reduce the manufacturing temperature including potential improvements in mix workability. The paper presents the experimental assessment in the form of the indentation test, compression strength test, 3-point bending beam strength test, stiffness and creep characteristics. The properties of mastic asphalt mixtures are complemented by parameters of the bituminous binders applied (complex modulus, viscosity etc.).

1 Introduction

Increasing loads on pavement structures have naturally emphasized not only in the Czech Republic but also in other developed countries the quality of asphalt mixtures used in pavement construction: ideally, these should be characterized by continuously improving durability and extension of pavement life-time. At the same time, pavement structures involve a number of critical aspects where the above mentioned is doubly true, mainly due to difficult pavement application or impossibility to properly compact a traditional asphalt formula. In many cases, traditional bituminous binders fail to meet the new requirements and their behavior and long-term performance must be modified by using suitable additives or polymers. In the field of road construction, this has been generally known for over 40 years. Nonetheless, the spectrum of options keeps expanding with the development of a range of new additives or variations on traditional modifiers. Modification of standard paving grade bituminous binders allows to improve the asphalt mix parameters with respect to its workability, load-bearing capacity and durability, or focusing primarily on individual characteristics in need of considerable enhancement (e.g. reducing occurrence of permanent deformation etc.).

S. El-Badawy and R. Abd El-Hakim (Eds.): GeoMEast 2018, SUCI, pp. 181–195, 2019.
https://doi.org/10.1007/978-3-030-01908-2_15

In order to reduce manufacturing temperatures, or improve workability, bituminous binders are modified by organic additives (synthetic waxes, fatty acid amides) using chemical additives (organo-silanes, amine or lignin based surfactants), or even mineral additives (zeolites). Except for the last instance mentioned, where the binder is modified by mineral substances inappropriate for MA technology due to the high temperature of the mixture and very fast vaporisation of the water bound in mineral additives (Ministry of Transport 2012; Valentin et al. 2013), modified binders can be applied to mastic asphalt layers to reduce manufacturing temperatures. We try to take advantage of the positives observed in cold asphalt mixes (Rondón-Quintana et al. 2015), as lower manufacturing temperatures and greenhouse gas reduction (Nazzal et al. 2011), even with respect to mastic asphalt mixtures where a manufacturing temperature lower by just 10–15 °C may reduce emission quantities by up to 30–50%. Partial results of Czech virtual centre for effective and sustainable transport infrastructure (CESTI) (Mondschein et al. 2016) point out the possibility of decreasing the manufacturing temperature to the 200 °C level while maintaining sufficient workability; however, mixtures with modified binders may demonstrate poorer stiffness. (Hofko et al. 2016) also mentions the possibility of reducing the manufacturing temperatures, improving deformation characteristics as well as a declared emission reduction by approximately 20% for mixtures prepared at 190 °C.

The experimental study completed by Czech Technical University in Prague (CTU) focused on the application of low-viscosity bituminous binders in mastic asphalt technology, particularly on modifiers and/or additives with a potential to affect the manufacturing temperature and, thus, the energy demands of the technology. In case of some alternatives the reachable benefits of chemical modification of bituminous binders by the low viscosity additives was evaluated as well. Due to that, in the experimental study the effect of bituminous binders modified by micromilled activated rubber powder (ARP) was analyzed as well, including the combination of the activated rubbed powder with low viscosity additives. The use of micromilled ARP is an alternative solution to crumb rubber modified bitumen in which the recycled material – waste rubber – is used in a non-standard form – smaller particles with very indented surface.

The issue of efficient waste rubber recycling has been closely followed for many years not only in the Czech Republic. In the road construction industry, different types of rubber granulate from old tires has been used to modify bituminous binders roughly since the 1980s. Nowadays, the crushed rubber granulate is subjected to the wet process which involves combining the bituminous binder, rubber granulate and possibly other agents, and such modified binder often called terminal blend is further applied to asphalt mixtures. In contrast to the application of industrial additives, which might increase the costs of pavement construction, bituminous binder modification using waste rubber is one of the efficient, economic and environmentally friendly options (Mashaan et al. 2013). The waste material is not only processed but the rubber addition even has a positive effect on the fatigue life and thermal induced cracking characteristics, prevention of crack propagation from underlying structural layers (bottom-up cracking), longer lifetime and, according to some sources, even noise emission reduction (Shu and Huang 2014; Ministry of Transport 2009). From our perspective, most of these benefits are true, only the last listed property requires a cautious approach. Ongoing long-term in situ noise measurements have not confirm unambiguously such benefit (Krivanek 2015, 2018). Certain negatives may be found in higher demand for energy, poorer mixture workability, storage and transport problems of the modified binder as the bituminous binder without any further modification must be used within a few hours from blending, otherwise segregates. This means the modification facility must often be located in immediate vicinity of the mixing plant, especially if bitumen is modified by more than 15% crumb rubber content (Fiedler et al. 2009). The presented study examined bituminous binders, which are modified by finely milled activated rubber wherein the granulated rubber is re-milled (to 0–6 mm or 0–4 mm) by high-speed milling technology with partial activation. The result is finely milled rubber powder (0–0.3 mm to 0–0.8 mm) which is used as a more effective way of bitumen modification.

The subject matter of the presented experimental study was the application of modified bituminous binders in mastic asphalt mixture MA8; the aim was to use such modified mastic asphalt for special applications like e.g. narrow pavement stripes close to bridge expansion joints (usually with a width of 0.5–1.0 m along the expansion joint). Although this application is the primary target, it certainly does not rule out the traditional application of modified mastic asphalt in larger areas of various types of structural pavement layers like bridge deck waterproofing layer or binder layer in case of two-layer bridge pavement structure.

With respect to the novelty of the assessed approach, use of modified bituminous binders in MA mixtures is very limited in Central Europe, use of crumb rubber modified binders has even in most European countries where mastic asphalt is used no tradition and only few experience. Typical crumb rubber was used in mastic asphalt mixtures in the past for some trial sections, but this approach failed mainly because of higher temperatures need for mixing the MA, increased odor, which was partly irritant to the workers and longer mixing times necessary to at least partly reach dissolution of the rubber particles in the bitumen. On the other hand, the use of synthetic waxes for mastic asphalt is a quite common approach; in Germany even mandatory if mastic asphalt mixtures shall be used. Experience with combining micromilled rubber particles with waxes for bitumen modification were not found in the literature.

2 Asphalt Mixture Design

For the study a suitable combination of crushed and freshly quarried aggregate was selected. The used aggregates hade to meet the requirements of technical standards from the perspective of granularity and, simultaneously, the mix design had to be economically acceptable from the perspective of the bituminous binder quantity added for the MA8-type of asphalt mixture. The selected mix version is marked FSV MA8 Valentin et al. (2016). The optimum quantity of bitumen was determined according to ČSN 73 6160. Generally the minimum allowed binder content for this type of asphalt mixtures is 7.0% by mass. The used optimum bitumen content for the presented mix design was experimentally verified to be 7.4% by mass (Table 1). This bitumen content was used for all variants. With respect to the previous findings concerning traditional compacted hot asphalt mixtures where rubber-modified binders demonstrated poorer workability at temperatures common for paving grade bituminous binders and polymer modified binders (due to its higher viscosity driven by the rubber particles in the binder), a version with an increased binder dose was tested as well in the case of modification by micromilled activated rubber (ARP), e.g. in Valentin et al. (2014a, b). The bitumen content was therefore increased by 0.5% which shall help to improve the workability of mastic asphalt where bitumen containing 10–15% micromilled rubber is used. The preparation of MA mixtures and requirements for used aggregate are defined in the standards ČSN EN 13043 and ČSN EN 13108-6.

Table 1. Mix design for the innovative mastic asphalt solution named "FSV MA8"

Mix	Bitumen content	Limestone filler content	Aggregate 4/8 mm (Zbraslav)	Sand 0/8 mm (Uhy)	Sand 0/4 mm (Uhy)
MA8 FSV (%)	7.40	22.24	42.12	3.70	24.54

3 Description of Bituminous Binders and Modifiers Applied

The reference mixture for the comparison of the particular innovative mix variants was a standard bituminous binder commonly used for mastic asphalts – hard paving grade bitumen according to ČSN EN 12591 marked as "20/30". Alternative solutions and the development of further improvement to the standard bituminous binder consisted of bitumen doping/modification by selected additives, which were based on new types of synthetic waxes, amide waxes (fatty amide acids) and active rubber powder from tire recycling processes. Mainly the last type of a modifier – pulverized rubber – was optimized and further developed during last five years in the Czech Republic. The activated micromilled rubber (ARP) or alternatively even vulcanizable activated rubber powder (VARP) with particle size of up to 0.8 mm was obtained by a special high-speed milling process like the process applied to Czech utility sample 29199. ARP was supplied by industrial study partner Lavaris s.r.o., which was involved in the development of the material alternative for a new type of bridge expansion joint damping profile throughout 2016 and 2017. Besides of the damping element the same activated rubber was used for bitumen modification. To improve ARP stability in the bituminous binder, a special butadiene-based activator was added to the overall composite as well. Table 2 gives the basic characteristics of bituminous binder designs with dopes or modifications as assessed below. Besides the conventional tests, the bituminous binders were also described by their penetration index which may provide not only an auxiliary material characterization but also information about its thermal susceptibility. To extend the knowledge of workability parameters, dynamic viscosity was determined which would give a hint of the future behavior of the mixture from the point of view of mixing or decreasing the manufacturing temperature in mastic asphalt preparation and application.

What is material from the point of view of penetration is the fact that the additives applied avoid significantly modifying this characteristic. The synthetic waxes applied

Table 2. Characteristics of bituminous binders used in MA mixtures

Bitumen	Penetration	Softening point	Penetration index	Dyn. viscosity @ 6.87 s^{-1}	
	(0.1 mm)	(°C)	(−)	135 °C (Pa.s)	150 °C (Pa.s)
20/30	20.5	62.8	−0.4	1.5	0.8
20/30 + 3% WE-CM	17.8	79.9	2.0	1.1	0.6
20/30 + 3% WE-BM	21.9	63.6	−0.1	1.0	0.7
20/30 + 15% CR (APR5AK)	15.0	81.3	1.8	13.8	12.4
20/30 + 15% CR + 2% WE-CM	14.2	87.9	2.5	13.8	11.6

increase the parameter slightly while ARP has a stiffening effect as is common in this type of modification. In the case of the softening point, the effect of synthetic wax WE-CM, resulting in a moderate increase of the characteristic, is of particular interest while the considerable softening point increase caused by ARP is of a fundamental significance. The dynamic viscosity as measured by rotational Brookfield viscometer is of no lesser importance. The selected temperature means that the effect of the waxes is minimal and the justification for wax application only becomes obvious from the point of view of the mortar in mastic asphalt. On the other hand, the application of micro-milled activated rubber powder results in an enormous viscosity increase as expected, which means deterioration of workability at the relevant temperature and, for the purpose of mastic asphalts, it indicates that the demand for mixing energy at standard manufacturing temperatures 200 °C might be higher than in the case of the reference bituminous binder. It is obvious that the application of wax in ARP-modified bituminous binder results then in a slight decrease of dynamic viscosity at 150 °C. There is no change in case of the 135 °C temperature.

4 Used Modifying or Improving Additives

4.1 Synthetic Waxes WE-CM and WE-BM

Bituminous binder modification by synthetic waxes is based on the need to modify the current bituminous binders in order to achieve a lower manufacturing temperature, or improve workability of the resulting asphalt mix (lower temperatures, easier application etc.). The mastic asphalt technology per se places a huge demand on energy (approx. 2.5 times more energy required than compacted asphalt mixes). Manufacturing temperatures of mastic asphalts reach up to 250 °C, so there is a need for economically more favorable solutions, for example in the form of bituminous binders modified by wax where even a minimum manufacturing temperature reduction by 10–15 °C already has a noticeable economical effect. The application of various types of waxes in mastic asphalt has been explored since late 1980's primarily with a view to improving asphalt mix workability (starting with Montana waxes, later continuing with Fisher-Tropsch waxes or fatty amide acids). In the last decade, it has been used as a tool to lower manufacturing temperatures and reduce emissions which might be released at higher temperatures.

The synthetic waxes used in the presented experimental study are innovative products not commonly used in road construction industry to improve bituminous binder properties yet. They have been sourced from a collaborating partner in Germany. Due to the need for intellectual property right protection it was not allowed to specify more relevant technical data and the chemical origin of these waxes. They are organic using different approach then, e.g. Fisher-Tropsch synthesis.

4.2 Active Rubber Powder (ARP)

Modification by active rubber powder has a relatively huge potential from both the economic and environmental perspectives as has been demonstrated (Soukupová 2014) already by the effort to use the material in bituminous binders intended for standard hot mix asphalts. With respect to the increasing requirements for non-renewable resource management and emphasis on waste material recycling at the same time, recycling of old tires (and not only those) for the purposes of bituminous binder modification presents an interesting technology. Naturally, we have to underline that from the industrial and social points of view, such activated material should be reused predominantly in the field of primary application – i.e. for rubber products. That is why the material was tested for instance for the purposes of manufacturing bridge expansion joints dampers where ARP was applied as a composite element compressed by elevated compression levels and at temperatures above 120 °C in the absence of any adhesives or sealants in 100%. At present, we see the added value of ARP modification of bituminous binders in traditional hot mix asphalts where the modification influences the rheological and mechano-physical properties of the final bituminous binder. Asphalt mixtures with modified bituminous binders demonstrate increased resistance to plastic deformation (improved stiffness of the asphalt mix) and improved resistance to thermal induced cracking (behaviour of the binder and, subsequently, the mixture in the low temperature range), see e.g. in (Valentin et al. 2014b). Some experts also declare a positive effect in the case of partial improvement to acoustic parameters (Kudrna et al. 2012) although no substantial noise dampening figures can be expected in this area.

The modified binder versions as described below work with bituminous binder modified by active rubber powder with a butadiene-based activator applied in the proportion of 5% of activated rubber content. The bituminous binder was modified by the so-called wet process where the activated, micromilled pulverized rubber is added to the heated bituminous binder (at least 170 °C), in our case in the quantity of 15% of the bituminous binder weight. The mixture was mixed in the laboratory at approx. 450 revolutions per minute for at least 30 min. High shear mixing was not applied in this case.

5 Mastic Asphalt Test Specimen Preparation

The preparation of the mastic asphalt (MA) mixture governed by ČSN EN 12697-35 +A1 defines the temperatures of individual mixture components and the method and duration of mixing. In our case, the manual mixing of the mixture is an important factor, which introduces the disproportionately longer mixing time and problems in maintaining the manufacturing temperature within the stipulated range. The preparation of test specimens (cubes for indentation test) as such followed ČSN EN 12697-20 in the case of the specimens for hardness (indentation) number and compression strength testing. The same method was applied to prepare cylindrical test specimens for the stiffness modulus test according to ČSN EN 12697-26 and cyclical compression test according to ČSN EN 12697-25 to determine deformation behaviour under controlled repeated stress. In the case of manufacturing beams for bending tensile strength testing, the possibility offered by ČSN 73 6160 to use reheated mixture was used. The beams (40 × 40 × 160 mm) were prepared with the mixture obtained by dissolving the specimens tested for the indentation number and compression strength test. It might be questionable if this reheating of already used mastic asphalt mixture is affecting the quality of material and therefore also the resulting properties. Since hard bituminous binders are used any ageing which occurs during reheating has neglecting impact on the mastic asphalt. Additionally it shall be pointed out that this approach is enabled by the standard and is used quite often. Finally, because all tested mixtures were reheated in the same way, if there would be a negative effect then it is leveled because of influencing all test specimens.

6 Workability

The test of workability of mastic asphalts is conducted according to ČSN 73 6160 and allows declaring any potential reduction of manufacturing temperatures. At the moment, the CTU Civil Engineering Faculty's road construction laboratory unfortunately lacks the equipment for the test, and the test was not performed even in the modified version thereof according to the national technical specification, TP 238. Therefore, the assessment of particular mix versions is currently just subjective, depending on the demands of manual mixing at the selected manufacturing temperature, and gives a merely informative indication. The designed mixture of aggregate and bituminous binder "FSV MA8" demonstrated subjectively poorer workability which was apparent in the mixing of the asphalt mixture (a need to keep mixing disproportionately longer, 10 min and more by way of example) while the standard mixing time should amount to roughly 5 min. The compacting in the moulds had to be hugely emphasized as well and some specimens had to be discarded due to excess air bubbles occurrence. From the point of view of subjective assessment of workability, mixtures with bituminous binder modified by active rubber powder had the poorest behavior as had been expected from the very beginning; it was also the reason behind the application of synthetic waxes in combination with ARP aiming to improve workability without affecting the properties of the mix.

7 Mastic Asphalt Mixture Parameters

The assessment of suitability of the modified binders for MA mixtures was made on basis of the indentation number, compression strength and bending tensile strength tests. The test set was also complemented by the findings obtained by stiffness modulus applying the test method set in EN 12697-26, annex C (IT-CY test method) and cyclical compression strength testing according to EN 12697-25.

7.1 Indentation Number

The basic parameter of mastic asphalt assessment is the determination of its indentation number on cube test specimens (hardness number). The way of test specimen preparation, specimen storage and the test method as such are described by ČSN EN 12697-20. The standard requires a (recommended) penetration depth of 1.0 to 5.0 mm in 30 min (for mix control testing it is between 0.6 mm and 6.5 mm) for mixtures using aggregate fractions 0/8 mm while the indentation number increase should not exceed 0.6 mm. The data presented does not comply with the marginal conditions in one segment thereof. The values obtained for mix versions FSV MA8 using improved or modified bituminous binders in 30 min are lower than the minimum indentation number 1.0 mm required for the mix design according to initial type testing (the aforementioned minimum threshold value is required with respect to reducing the risk of excess asphalt mix brittleness and crack prevention). In the case of applying the criteria for control tests, versions modified by micromilled activated rubber fail to pass. Despite the results, certain obvious dependencies on the choice of bituminous binder applied and the potential benefit of individual bituminous binder modifications or improvements are observed. The mixtures modified by active rubber powder (Fig. 1, mix variant 20/30 + CR) demonstrate a significant improvement to the indentation number with an approx. 60% drop relative to the reference mixture. Modification by synthetic organic waxes also contributes towards lowering the indentation number, and the best results are achieved by a bituminous binder modified by a combination of active rubber powder (ARP) with synthetic wax.

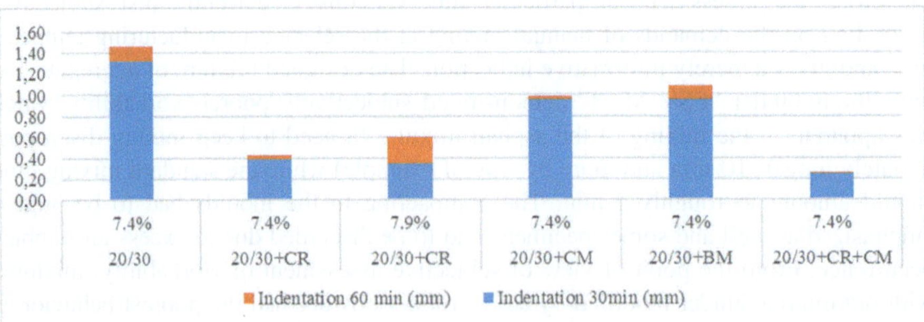

Fig. 1. Results of the indentation test for assessed FSV MA8 mix variants (y-axis: indentation number in mm)

7.2 Compression Strength

Another parameter in the assessment of modified binder suitability for mastic asphalt layers is its compression strength tested on cubes. Compression strength was measured both in standard dimension cubes, with width and height a = 70 mm, and in cubes prepared with the material left over after the bending tensile strength test (a = 40 mm). The absolute values from the two sets of measurements differ to a certain degree; nevertheless, the trend for comparing the individual bituminous binders is similar in both cases, see Fig. 2. Relative to the reference mix, the versions with improved or modified binders achieve considerably higher compression strengths. Again, the hypothetical winner is the bituminous binder modified by a combination of active rubber powder and on type of the tested synthetic wax representatives. Even in this case, the binder modification had a positive effect on the parameters assessed (see Table 3). The recommendation from the existing Czech standards according to which the resulting compression strength should range between 3.5 MPa and 8.0 MPa is challenged here. Even in this case, the threshold value is stipulated in order to avoid excess stiffness of the material, which is usually accompanied by high level of brittleness and increased risk of cracking in the low temperature range (thermal induced cracking). Nevertheless it is necessary to accentuate that the compression strength range as is defined for mastic asphalt mixtures in the Czech Republic results from the assumption that hard penetration grade bitumen, eventually standard 50/70 paving grade are used. In the presented experimental study the binders were modified by crumb rubber which seems to have a significant effect on increased value of the strength property. Because of the elastic characteristics a rubber naturally has this does not necessarily mean that higher compression strength values will increase the risk of cracking. For mastic asphalt mixtures containing CRMB binders it would be most probably necessary to increase the upper limit of the acceptable compression strength (Table 3).

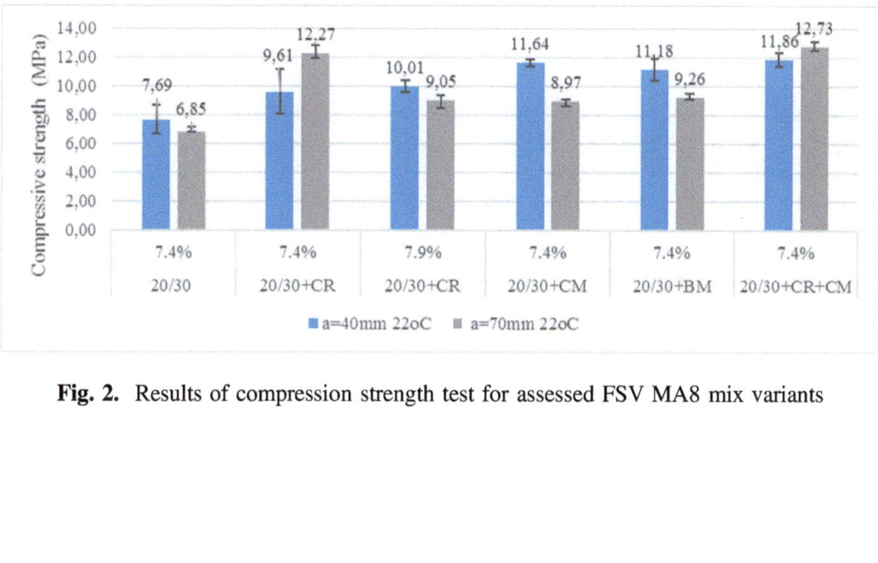

Fig. 2. Results of compression strength test for assessed FSV MA8 mix variants

Table 3. Selected mix characteristics for the compared MA8 variants

Bitumen	Binder content	Bending tensile strength	Modulus of elasticity	Compression strength @ 22 °C		Indentation number after 30 min	Increment in indentation number after 60 min
		22 °C	22 °C	a = 40 mm	a = 70 mm		
		(MPa)	(MPa)	(MPa)	(MPa)	(mm)	(mm)
20/30	7.4%	6.83	245	7.69	6.85	1.33	0.14
20/30 + CR	7.4%	6.78	340	9.61	12.27	0.39	0.04
20/30 + CR	7.9%	9.35	519	10.01	9.05	0.35	0.25
20/30 + CM	7.4%	11.05	485	11.64	8.97	0.95	0.03
20/30 + BM	7.4%	10.74	433	11.18	9.26	0.95	0.14
20/30 + CR + CM	7.4%	6.33	630	11.86	12.73	0.25	0.02

7.3 Bending Tensile Strength

The assessment of bending tensile strength, or the characteristic calculated therefrom – the modulus of elasticity - challenges the limits stipulated by the applicable national technical standard where the mixtures with wax-modified binders exceed the maximum recommended value of 7.0 MPa (see Fig. 3). Disregarding this fact, the modification or improvement of bituminous binders facilitated an improvement in the compression strength parameters, and even the modulus of elasticity increased relative to the reference mix. In this case, the best performers were mixtures with bituminous binder modified by innovative waxes WE. Even the version of mastic asphalt mix with a higher quantity of bituminous binder modified by micromilled ARP demonstrated a higher value. Both mixtures show approximately 40% increase in the bending tensile strength parameter.

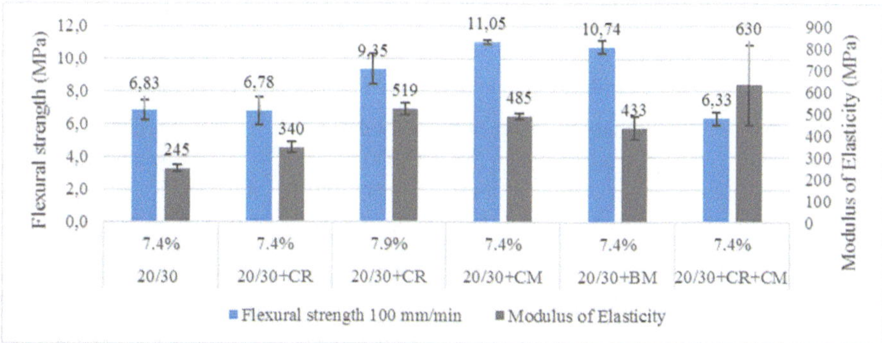

Fig. 3. Results of bending tensile strength test and the modulus of elasticity for assessed FSV MA8 mix variants

It is just necessary to carefully consider the shown result of modulus of elasticity for the mix variant 20/30 + CR + CM. The error bar in this case sets a range which is almost 50% of the measured value. Additionally test replicates would be necessary to confirm the presented results of this mix variant.

Similarly to the compression strength values it is questionable for the future if due to presence of crumb rubber modified bitumen and its improved elastic behavior the actually applied threshold values have not to be tilted or if mastic asphalts containing modified binders should not have a separate quality category to differentiate them from traditional mastic asphalt with either hard penetration grades (e.g. 20/30) or regular paving grade bituminous binders (e.g. 50/70).

7.4 Asphalt Stiffness

The test method for the determination of stiffness modulus is described in ČSN EN 12 697-26. As already mentioned earlier test method C was applied, i.e. non-destructive test of repeated tensile stress at different temperature. Generally, it has been pointed out that the determination of stiffness modulus is not a traditionally measured characteristic for mastic asphalt mixtures. In the case of stiffness moduli, there is an obvious positive effect of bituminous binders modified by active rubber powder as summarized in Table 4. The test specimens demonstrated noticeably superior results when the stiffness was measured at higher temperatures – which potentially indicates their suitability for applications aiming to reduce the permanent deformation occurrence (rutting). The thermal susceptibility of the mixture is a very interesting parameter; the mixtures showed a substantial decrease in thermal susceptibility relative to the reference mix. In this case, bituminous binder modification by synthetic waxes translated to a rather negative effect with respect to stiffness moduli measured at higher temperatures (27 °C and 40 °C), and to inferior thermal susceptibility. When the bituminous binder was modified by synthetic wax WE-CM, the higher thermal susceptibility stemmed from the lower stiffness modulus value at 40 °C, which is the basis for thermal susceptibility assessment. From the perspective of stiffness testing at higher temperatures, the values were comparable to the reference mixture.

Table 4. Results of stiffness determination for assessed FSV MA8 mix variants

Bitumen	Binder content	Stiffness modulus "S" @T = (MPa)				Thermal susceptibility	
		0 °C	15 °C	27 °C	40 °C		
		(MPa)	(MPa)	(MPa)	(MPa)	S_0/S_{27}	S_0/S_{40}
20/30	7.4%	19157	10247	5155	2124	3.7	9.0
20/30 + CR	7.4%	19344	13101	8500	5037	2.3	3.8
20/30 + CR	7.9%	17597	11738	6612	2998	2.7	5.9
20/30 + CM	7.4%	19504	10047	4406	1553	4.4	12.6
20/30 + BM	7.4%	20113	10776	5507	2139	3.7	9.4
20/30 + CR + CM	7.4%	19504	14857	8363	3826	2.3	5.1

7.5 Cyclical Compression Test

The test method described in ČSN EN 12697-25 determines the resistance of the mixture to permanent deformation caused by cyclical stress on cylindrical test specimens at elevated temperatures. In the case of mastic asphalt, the value of permanent deformation according to the aforementioned test can be also expressed by the dynamic indentation number. The results measured correspond with the indentation number measurements where the dynamic indentation number achieves significantly superior values in MA mixtures with bituminous binders modified by active rubber powder (see Fig. 4). The test shows that the use of waxes in the mastic asphalt mixture does not modify the characteristics on test while, contrastingly, used CRMB binder or a combination of micromilled activated rubber and synthetic wax increases the characteristic significantly in the case of creep modulus.

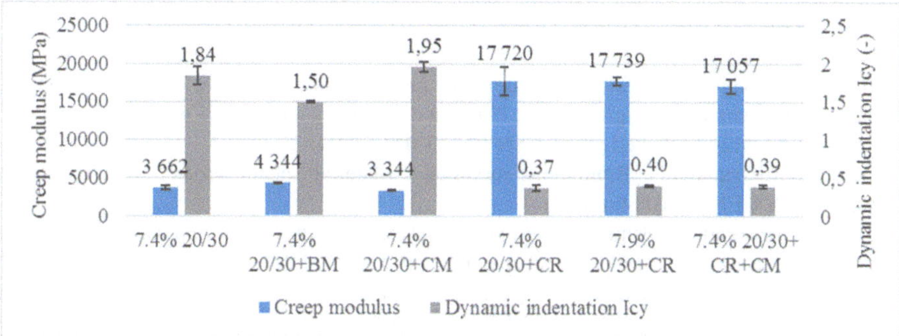

Fig. 4. Graphic visualization of creep modulus and dynamic indentation number from the cyclic compression test (at T = 60 °C) for assessed FSV MA8 mix variants

8 Conclusions

The versions of the new MA8 mix designs with modified or improved bituminous binders on test demonstrated an interesting potential where the addition of the assessed modifiers results in a positive effect on the final asphalt mix characteristics or performance based parameters. The positive contribution of bituminous binders modified by micromilled active rubber was clearly proven when mixtures with the modified bitumen demonstrated an increase in strength characteristics, low thermal susceptibility (as shown in Table 4), high stiffness modulus even at higher temperatures, high resistance to cyclical stress and permanent deformation. Unfortunately, the improvement to the parameters was offset by subjectively poorer workability where the mixtures presented higher demand for mixing energy during test specimen preparation. Bituminous binder modification by the new generation of synthetic waxes, WE-CM and WE-BM, also demonstrated a benefit in the form of increased strength parameters, particularly bending tensile strength. From the perspective of the stiffness modulus, thermal susceptibility and resistance to permanent deformation on test, bituminous binder modified by wax WE-BM shows more potential from the point of view of mastic asphalt as the asphalt mixtures with this binder achieved the same or slightly superior parameters relative to the reference mix. The level of effect the waxes exert on the workability and thus the potential manufacturing temperature decrease has not been verified or sufficiently proven yet, and will remain the subject of further research.

Acknowledgements. This paper was elaborated within the activities of the Centre for effective and sustainable transport infrastructure (CESTI) - TE01020168 as supported by the Technology Agency of Czech Republic.

References

Czech office for standards: Testing of bituminous mixtures for roads ČSN 73 6160 (2008)

Czech office for standards: Aggregates for bituminous mixtures and surface treatments for roads, airfields and other trafficked areas. ČSN EN 13 043 (2004)

Czech office for standards: Bituminous mixtures - Material specifications - Part 6: Mastic Asphalt. ČSN EN 13 108-6 (2008)

Czech office for standards: Bituminous mixtures – Test methods for hot mix asphalt – Part 20: Indentation using cube or cylindrical specimens (CY). ČSN EN 12 697-20 (2012)

Czech office for standards: Bituminous mixtures - Test methods for hot mix asphalt - Part 25: Cyclic compression test. ČSN EN 12 697-25 (2006)

Czech office for standards: Bituminous mixtures – Test methods for hot mix asphalt – Part 26: Stiffness. ČSN EN 12 697-26 (2012)

Czech office for standards: Bituminous mixtures - Test methods for hot mix asphalt - Part 35: Laboratory mixing. ČSN EN 12 697-35+A1 (2007)

Fiedler, J., Komínek, Z., Bureš, P., Kašpar, J., Racek, I.: Využití pryžového granulátu z pneumatik do vozovek 14. Ročník konference Recycling 2009, pp 30–34 (2009)

Hofko, B., Dimitrov, M., Schwab, O., Weiss, F., Rechberger, H., Grothe, H.: Technological and environmental performance of temperature-reduced mastic asphalt mixtures. Road Mater. Pavement Des. **18**, 22–37 (2016)

Krivanek, V., et al.: Change of road surface noisiness during several-year usage. Final report, Brno, Transport Research Centre. 88 pages. Sponsor: Technology Agency CR. (only in Czech: Změna hluku povrchů vozovek v průběhu několika let používání) (2015)

Krivanek, V., et al.: Tools for Analysis and Assessment of Environmental Impacts of Road Surface Noise. Final report (summary), Brno, Transport Research Centre. 12 pages. Sponsor: Technology Agency CR. (only in Czech: Nástroje pro analýzu a hodnocení environmentálních dopadů hluku vozovek) (2018)

Kudrna, J., Dasek, O., Kachtík, J., Urbanec, K.: Asfaltové vrstvy s gumoasfaltovým pojivem. In: Stavitel 2012(11), 2, Economia a.s., Prague. (only in Czech: Asphalt layers with crumb-rubber modified bituminous binders) (2012)

Mashaan, N., Ali, A.H., Koting, S., Karim, M.: Performance evaluation of crumb rubber modified stone mastic asphalt pavement in Malaysia Hindawi Publishing Corporation. Adv. Mater. Sci. Eng. **2013**, 8 pages (2013). Article ID 304676, https://doi.org/10.1155/2013/304676

Ministry of Transport - Road Infrastructure Department.: Hutněné asfaltové vrstvy s asfaltem modifikovaným pryžovým granulátem z pneumatik TP 148 (2009)

Ministry of Transport - Road Infrastructure Department.: Nízkoteplotní asfaltové směsi (NTAS) TP 238 (2012)

Mondschein, P., Miláčková, K., Soukupová, L., Valentin, J., Bureš, P., Kašpar, J.: Laboratory assessment of new technical solutions for mastic asphalt with reduced mixing temperature. In: 6th Eurasphalt and Eurobitume Congress (2016)

Nazzal, M.D., Sargand, S., Al-Rawashdeh, A.: Evaluation of warm mix asphalt mixtures containing RAP using accelerated loading tests. J. Test. Eval. **39**(3), 1–8 (2011)

Rondón-Quintana, H.A., Hernández-Noguera, J.A., Reyes-Lizcano, J.A.: A review of warm mix asphalt technology: technical, economical and enviromental aspects. Ingeniería e Investigación **35**(3), 5–18 (2015)

Shu, X., Huang, B.: Recycling of waste tire rubber in asphalt and Portland cement concrete. Overv. Const. Build. Mater. **67**(Part B), 217–224 (2014)

Soukupová, L.: Functional Characteristics of Bituminous Binders with Micro-Ground Rubber and Various Types of Catalysts and Chemical Additives. Master thesis, Czech Technical University in Prague (2014)

Valentin, J., Mondschein, P., Valentová, T.: Ověření a podmínky aplikace minerálního materiálu pro NTAS s obdobnými vlastnostmi jako zeolity Technické listy Centre for effective and sustainable transport infrastructure (CESTI) (2013)

Valentin, J., Mondschein, P., Kolařík, J., Bureš, P.: Soubor ověřených technických parametrů nízkoteplotních litých asfaltů pro rozšíření technických podmínek TP 238. Výzkumná zpráva DVZ 1.1.2, FSv ČVUT v Praze (2014)

Valentin, J., Mondschein, P., Šedina, J., Benešová, L., Landa, F.: Lité asfalty s netradičními pojivy. XXI. Seminar I. Poliačka, Bratislava (2016)

Valentin, J., et al.: Characteristics of selected bituminous binders with high-speed grinded crumb rubber. In: Transportation Research Arena 2014. Transportation Research Arena, Paris, 11–14 April 2014, pp. 1–10. IFSTTAR, Marne-la-Vallée (2014a)

Valentin, J., et al.: Durable acoustic asphalt pavement courses with utilization of bituminous binders modified by rubber microfiller including innovative technology of rubber milling. Final report, Prague, Transport Research Centre. 79 pages. Sponsor: Technology Agency CR (only in Czech) (2014b)

Retraction Note to: Improvement of Rheological Properties of Asphalt Binder by Adding Composite Montmorillonite Nanoclay

Abdulhaq Hadi Abedali Al-haddad(✉) and Rana Amir Yousif

Retraction Note to: Chapter "Improvement of Rheological Properties of Asphalt Binder by Adding Composite Montmorillonite Nanoclay" in: S. El-Badawy and R. Abd El-Hakim (Eds.): GeoMEast 2018, SUCI, https://doi.org/10.1007/978-3-030-01908-2_4

The Editors have retracted this conference paper because it significantly overlaps with previously published articles by different authors [1, 2].

The authors disagree with this retraction.

[1] R. Muniandy et al., Effect of Organic Montmorillonite Nanoclay Concentration on the Physical and Rheological Properties of Asphalt Binder, Aust. J. Basic & Appl. Sci., 7(9): 429-437, 2013

[2] R. Muniandy et al. Influence of Clay Nanoparticles on the Physical and Rheological Characteristics of Short Term Aged Asphalt Binder, Australian Journal of Basic and Applied Sciences, 7(11) Sep 2013, Pages: 488-498

The retracted version of this chapter can be found at https://doi.org/10.1007/978-3-030-01908-2_4

S. El-Badawy and R. Abd El-Hakim (Eds.): GeoMEast 2018, SUCI, p. C1, 2025.
https://doi.org/10.1007/978-3-030-01908-2_16

Author Index

S. El-Badawy and R. Abd El-Hakim (Eds.): GeoMEast 2018, SUCI, p. 197, 2019.
https://doi.org/10.1007/978-3-030-01908-2

MIX
Papier aus verantwortungsvollen Quellen
Paper from responsible sources
FSC® C105338

If you have any concerns about our products,
you can contact us on
ProductSafety@springernature.com

In case Publisher is established outside the EU,
the EU authorized representative is:
Springer Nature Customer Service Center GmbH
Europaplatz 3, 69115 Heidelberg, Germany

Printed by Libri Plureos GmbH
in Hamburg, Germany